3195075

Paleobiogeography

Topics in Geobiology

Series Editor: Neil H. Landman, American Museum of Natural History, New York, New York
Douglas S. Jones, University of Florida, Gainesville, Florida

Current volumes in this series

Paleobiogeography

Bruce S. Lieberman

University of Kansas
Lawrence, Kansas

Kluwer Academic / Plenum Publishers
New York, Boston, Dordrecht, London, Moscow

Library of Congress Cataloging-in-Publication Data

Lieberman, Bruce S.
    Paleobiogeography: Using fossil to study global change,
plate tectonics, and evolution/Bruce S. Lieberman
       p.   cm. — (Topics in geobiology; v. 16)
    Includes bibliographical references and index.
    ISBN 0-306-46277-X
      1. Paleobiogeography. I. Title. II. Series
QE721.2.P24 L54 1999
560—dc21   99-055322

ISBN: 0-306-46277-X

© 2000 Kluwer Academic / Plenum Publishers, New York
233 Spring Street, New York, New York 10013

http://www.wkap.nl/

10 9 8 7 6 5 4 3 2 1

A C.I.P. record for this book is available from the Library of Congress.

Printed in the United States of America

For Paulyn and Sarah

Foreword
by Daniel R. Brooks

This book represents the first major attempt to integrate the "space, time, and form" of historical biogeographers with the "tempo and mode" of paleontologists. The author is young, as befits a pioneer, but he is uniquely qualified for this undertaking, having been successively a student of Steve Gould, Niles Eldredge, and Elisabeth Vrba (and now plying his trade at the same institution that houses Charles Michener and Ed Wiley). His published repertoire includes theoretical studies on macroevolution, punctuated equilibrium, and historical biogeography, as well as monographic studies on the taxonomy, systematics, and phylogeny of trilobites.

Everything we have learned about biological evolution, and everything we hope to learn about it, is based on the recognition that the past leaves its mark on the present, and does so in such a striking way that we can actually understand at least part of what happened in the past by finding these unique historical marks in the present. Patterns of inheritance from one generation of fruit flies to the next in a series of glass vials in a laboratory are just as historical as the sequence of fossils in a rock outcropping exposed by the cutting action of running water. Long before the advent of modern evolutionary thinking, naturalists were impressed by the fact that biological diversity occurred nonrandomly across the Earth. In fact, a hallmark of a great naturalist is knowing where to find things. Thus it is no surprise that the first two major lines of evolutionary thought, Lamarckism and Darwinism, reserved a special explanatory role for the relationship between geography and evolution. Lamarck felt that wherever a species happened to evolve, it would deal with changes in the local environment by evolving the appropriate adaptations (in his wonderful Panglossian world, nothing ever went extinct). Darwin, on the other hand, felt that species would deal with changes in their local environments to the best extent possible given the constraints of their inherited variation, and might survive or might go extinct. In both cases, the historical record left by the survivors should, among other things, provide a strong imprint of a causal relation between Earth history and biological history. Biogeographic studies have thus been a major part of evolutionary biology from the beginning and remain a critical cornerstone today. With the advent of evolutionary thinking, paleontologists realized fully that where could also mean when. They became the curators of the "evolutionary epic" of evolved biological diversity deployed across space and through time.

The mid-1960s to mid-1970s witnessed a revolution in paleontology, initiated by the theory of punctuated equilibrium proposed in 1972 by Niles Eldredge and Steve Gould. This rapidly led to the elaboration of a hierarchical view of evolution, promoted most strongly by Eldredge, Gould, and Elisabeth Vrba, that included a renewed appreciation for the reality of macroevolutionary phenomena and the role of ontogenetic constraints in evolutionary diversification. The emerging field of paleoecology, championed by such luminaries as David Raup, John Sepkoski, Art Boucot, and Dave Jablonski, produced significant insights into the nature of biotic turnover and replacement. This has led to a widespread belief, synthesized by Joel Cracraft in 1985, that geological evolution has played a major role in determining the rate of diversification of life on this planet, through speciation mediated by fragmentation of ancestral species owing to geological evolution and extinction owing to changes in environmental harshness on regional to global scales. Studies by the past generation of paleontologists confirm that the deployment of biodiversity through time involved episodes of biotic expansion and habitat fragmentation leading to speciation alternating with episodes of regional to global-scale extinction.

At the same time as this new paleontology was emerging, three events energized biogeography, the study of the deployment of biodiversity in space. The first was the publication of the theory of island biogeography, which provided an analytical and quantitative basis for ecological biogeography, a discipline rich in tradition and narrative accounts about the relationships between current environments and the number and relative abundance of species in particular areas. The second was the acceptance of the theory of plate tectonics and continental drift by geologists. This breakthrough permitted biologists to recapture a wealth of ideas linking geographic distributions of related species to general distribution patterns on a global scale. From 1890 to 1940 general biogeographic distribution patterns were explained by reference to the theory of continental drift. Faced with hostility from the geological community, however, historical biogeographers abandoned continental drift for land bridges or chance dispersal across fixed geographical barriers mediated by climate. The third was the linking of the methods of phylogenetic systematics developed by Willi Hennig to biogeogrpahic studies.

A seminal study by Lars Brundin provided the first modern evidence of what we now explain as trans-Gondwanian distribution patterns mirrored in the distributions of extant taxa. Gareth Nelson wrote an article in 1969 on "the problem of historical biogeography," noting that if one estimated ancestral distributions by adding together the distributions of sister species and designating the combined distribution as the ancestral distribution at the node on the phylogenetic tree linking those sister species, the implied geographic distributions indicated great antiquity for many modern groups. Was it possible that historically associated remnants of ancient biotas existed in geographic areas that are now disjunct but which were connected in the distant past?

The merger of phylogenetic systematics and the new paleontology has been exciting. As anticipated in the early 1970s by noted systematists such as Eldredge, Wiley, and Colin Patterson, fossils have been willing to give up their secrets as readily as specimens representing living species. That fossils provide data critical for the resolution of key aspects of phylogenetic analysis including living species has been amply demonstrated by systematists such as Michael Donoghue, Jacques Gauthier, Arnold Kluge, and Michael Novacek. Phylogenetic hypotheses based on fossil species have been used to evaluate hypotheses of the age of origin and diversification for particular taxa and to address the question of whether or not diversification and extinction are coupled and/or episodic. These studies have shown that there is a great deal of concordance between the temporal dimension discovered by paleontologists and the morphospace dimension inferred from phylogenetic analysis. In fact, this concordance is strong enough that molecular evolutionary biologists use the fossil record to calibrate statistical models of evolutionary diversification at the level of nucelotide sequences (the "molecular clocks").

Interestingly, the Hennigian revolution has had, until now, relatively little impact on paleontological studies of biogeography, despite the recognition that both space and time are important parameters in evolution. This is likely due to the belief that geographic distributions for fossil taxa are too fragmentary for adequate study, despite the existence of extremely dense fossil records for a number of taxa. In fact, my first contact with the author of this book occurred after a major systematic biology journal had rejected one of his manuscripts on the grounds that historical biogeographical studies using fossils were impossible. Fortunately, the editor and reviewers of *Paleobiology* were more enlightened and the work was published. It was a seminal paper.

Early attempts to incorporate the possibility of episodes of biotic expansion and fragmentation into phylogenetic biogeographic studies presumed that the common ancestor of a clade was widely distributed throughout all the areas in which subsequent vicariant speciation occurred. In the early 1990s, Kare Bremer developed "ancestral area analysis" to initiate a more realistic assessment of ancient biotic distributions, but this remains controversial among neontologists. This is clearly a critical area for the development of historical biogeography, and one in which paleontology should make a unique contribution. The author of this book is a paleontologist who has set out to make that contribution. He has proposed a method that permits independent assessment of both vicariance, or speciation owing to habitat fragmentation, and geodispersal, or fusing of areas due to sea-level drops and/or tectonic fusion, allowing biotas to expand into new territory. Comparison of these assessments permits us to "see" the episodic expansion and fragmentation of biotas and to assess the extent to which the former might have set the stage directly for the latter.

This book will not be the last word on this subject, not even by the author himself, for it sets out a way to begin an entirely new kind of research and to use the results in integrated explanations of evolution. That is what makes it an exciting frontier in evolutionary biology. Enjoy!

Preface

Among the topics that I will consider in this book are: (1) why do different regions have markedly different floras and faunas (both fossil and living); (2) why, when we study the fossil record do we find that at different times in the history of life there seems to be a greater or lesser number of regions with largely unique floras and faunas; (3) do closely related groups of organisms tend to occupy the same regions; and (4) what are the various ways to analyze the differences between different floras and faunas? These are actually a subset of the major questions that have been posed throughout the years by biogeographers. Some additional questions have been nicely encapsulated by Brown and Lomolino (1998, pp. 3–4), and it is worthwhile to summarize some of them here. Specifically, (5) why is a species or higher taxon confined to its present range; (6) what enables a species to live where it does and prevents it from colonizing other areas; (7) what are the roles of climate, barriers, and competition in controlling the distribution of species; (8) where do two closely related species occur in relation to one another; (9) why do large isolated regions possess very distinct animals and plants; (10) why are there more species in the tropics than near the poles; and (11) how are islands colonized?

This book is organized into a set of chapters introducing themes that will enable the reader to visualize the field of paleobiogeography from several different angles. Chapter 2 focuses specifically on the relationship between hierarchy theory and biogeography. Here I discuss why there is a distinction between what has been referred to as phylogenetic or historical biogeography (*sensu* Brooks and McLennan, 1991), essentially the subdiscipline of biogeography that attempts to elucidate evolutionary patterns and processes and their relationship to the Earth's geography, and ecological biogeography (*sensu* Brown and Lomolino, 1998), essentially the subdiscipline of biogeography that attempts to elucidate ecological patterns and processes and their relationship to the Earth's geography. Chapter 3 is a discussion of which types of biogeographic patterns are most amenable to study in the fossil record; i.e. those patterns that are of the greatest relevance to paleobiogeographers, and also includes a discussion of the quality and limits of resolution of the fossil record.

Chapter 4 considers the historical development of the field. Many of the current major debates in the field of biogeography have a long intellectual tradition, and our understanding of the issues will increase when we consider them from an historical perspective. Chapters 5 and 6 consider revolutions in the fields of geology and biology, and their respective impact on biogeography and paleobiogeography. In Chapter 7, I discuss how biogeographic areas are defined. Chapters 8 and 9 deal with the analysis of paleobiogeographic data, including case studies to clarify the techniques. Finally, in Chapter 10, I conclude with a commentary on the significance of biogeography and paleobiogeography for our understanding of the current biodiversity crisis and the associated mass extinction that afflicts the flora and fauna of this planet. In each of these thematic chapters, the role of paleobiogeography as a bridge between the diverse fields of biology and geology is emphasized.

As an inspiration to those embarking on reading this book, let me quote the opening sentence of Charles Darwin's *On the Origin of Species*...

> When on board H.M.S. *Beagle*, as naturalist, I was much struck with certain facts in the distribution of the organic beings inhabiting South America, and in the geological relations of the present to the past inhabitants of that continent. These facts, as will be seen in the latter chapters of this volume, seemed to throw some light on the origin of species—that mystery of mysteries, as it has been called. (Darwin, 1872, p. 27)

Acknowledgments

There are many people who have helped make this book possible. My editor, Ken Howell, was extremely helpful and very supportive of this project, and I enjoyed working with him very much. His assistant, Arne Olsen, also was very helpful. I am also grateful to the scientific editors of the Topics in Geobiology series and to Kluwer Academic/Plenum Publishers for believing in this project. Throughout my work on this book my family was very understanding and supportive and for that I am very appreciative. I thank Dan Brooks for numerous discussions and also for his theoretical insights into biogeography. Niles Eldredge was a role model. Tim White provided significant help with various logistical matters. The faculty in the Departments of Geology and Ecology and Evolutionary Biology at the University of Kansas also helped facilitate the completion of this project. Among these individuals Roger Kaesler, Bob Goldstein, Dick Robison, Bert Rowell, Don Steeples, Tony Walton, Tom and Edie Taylor, and Ed Wiley deserve special mention. NSF EAR-9505216 and the University of Kansas provided financial support. Finally, the genius, professional conduct, and toil in the field of Alfred Russell Wallace was inspirational.

Contents

Paleobiogeography

Chapter 1

What Is Paleobiogeography?

The field of paleobiogeography is derived in direct lineage from its close intellectual cousin, biogeography, a scientific discipline that had its birth in the adventurous and exploratory urges of humankind. When people traveled to far-off lands centuries ago they discovered, much to their surprise, that the plants and animals of these lands, as well as the fossils preserved in their strata, were often very different from those they were familiar with in their own countries. Indeed, it "...might have been anticipated (that there would be) an almost perfect identity in the animals and plants which inhabit corresponding parallels of latitude" (Lyell, 1832, p. 66), i.e., similar climates. Instead, the travelers' senses were enriched with the sounds of new bird songs, the smells of new plants, and the sights of different peoples and cultures. The motivations for these explorations were often both base and noble: the will to power or to dominate a local people by controlling their resources, but also the desire to know and to understand. As this is a scientific text, rather than a sociological one, I am going to focus on what are best perceived as the more noble intentions of humanity. Specifically, how the quest for knowledge about our world is related to the establishment of a tradition in paleobiogeographical and biogeographical research.

Exploration has a double-edged meaning in biogeography. We are trying to use a scientific framework as an exploratory tool to understand the world around us, but we are also relying on explorations in other parts of the globe as the database from which all of our analyses derive. Biogeography is a discipline that seeks to explore what the geographical distribution of organisms can tell us about the relationship between the evolution of the Earth's biota and the evolution of the Earth itself. This is a question that has intrigued scientists for centuries. Indeed, some scientists have argued that there is "one great thought (that) prevails in natural historical studies, the study of the laws regulating the geographical distribution of natural families of animals and plants upon the whole surface of the globe" (Agassiz, 1842, quoted in Browne, 1983, p. 138). This is because the "Earth and life evolved together" (Croizat, 1964, p. 46).

It is necessary to delineate at the outset what exactly is meant by paleobiogeography. The reader will appreciate the obvious concatenation in the word "paleobiogeography" of the prefix "paleo" with the suffix "biogeography", signifying some intersection between old or ancient and

1

biogeography. In truth, here old or ancient really means that the fossil record is the venue for paleobiogeography—the data bearing on problems in paleobiogeography are fossils. The strengths and weaknesses of the field will depend on the inherent strengths and limitations of the fossil record, but etymologically, as well as scientifically, the field is bound to the discipline of biogeography. While the term paleontology seems crystal clear, perhaps the meaning of biogeography is not so straightforward.

When we break the word biogeography down into its constituent parts, we find much that is useful. First we see an association between life and geography, which is in essence what the subject is all about. Biogeography is the study of the distribution of different types of animals and plants, and the biological, geological, and climatic processes that influence this distribution (Brown and Lomolino, 1998). At its core, biogeography is the discipline dedicated to elucidating how the Earth and its biota, its complement of animals and plants, have evolved and coevolved (Brooks, 1981, 1985; Brooks and McLennan, 1991; Brooks *et al.*, 1981; Croizat *et al.*, 1974; Nelson, 1978; Platnick and Nelson, 1978; Wiley, 1988*a,b*). Biogeography is a science that attempts to reconstruct how organisms are distributed over the present surface of the Earth (Brown and Lomolino, 1998), and paleobiogeography adds in the element of the history of the Earth.

Paleobiogeography involves the study of patterns in the history of life. However, it also provides information relevant for the study of evolutionary, ecological, and geological processes (Brooks and McLennan, 1991; Brown and Lomolino, 1998) because "the most important connection between the two areas (pattern and process) involves the comparison of the patterns of both intrinsic and *extrinsic* features of organisms predicted from theories of process, with those actually 'found' in nature" (Eldredge and Cracraft, 1980, p. 4). The types of evolutionary processes that can be studied using paleobiogeographical and biogeographical data include determining how the dissimilarities between animals and plants in different regions come about and the dominant manner in which the speciation process works? The types of ecological processes that we can study include the mechanisms that regulate diversity in a particular region and the mechanisms that control the presence or absence of a taxon in a region. In addition to these evolutionary and ecological processes, we can also study various geological processes, largely manifestations of plate tectonics and climatic change, and their role in accelerating or retarding evolutionary change.

Obviously, studies that consider biogeographic patterns in the extant biota have certain advantages over paleobiogeographic studies, which deal with biogeographic patterns in the fossil record. First, researchers can use molecular markers to study the relationships between extant species and use this information to infer biogeographic patterns. They are also able to manipulate systems experimentally and study biogeographic patterns on very small spatial and temporal scales. Further, when a biogeographer samples a modern biota for its diversity, he or she is likely to obtain a more accurate picture of its true diversity than a paleobiogeographer studying a fossil biota. There is one crucial area, however, where studies of the extant realm do not, and cannot, surpass those of the fossil realm, and that is in regard to timescale. Studies of the modern biota are limited to a single temporal interval. We can use these studies to infer something about the bioa's history, but we cannot track that history.

A modern biota may have been affected by several cycles of biogeographic change, and remnants or residues of many of these cycles will be confusingly effaced or jumbled. The most recent cycle may have strongly overprinted previous cycles, or it might have left them unperturbed and thus all the more difficult to interpret. This is why study of the modern biota alone cannot provide a complete understanding of biogeographic patterns and processes. For such an understanding, we also need the fossil record, which is our one true chronicle of the history of life. Regardless of how complete or incomplete the fossil record is, if we wish to reconstruct that history as it unfolded, and not view it as a static final chapter, then we need to look at that record.

In the fossil record, paleobiogeographers can actually monitor how the Earth and its biota have evolved and coevolved. They can look at the association between major geological events and biogeographic patterns over long timescales, through several time slices. Biogeographers have access to only one picture of diversity, but that single picture may be exceedingly clear. Paleobiogeographers, on the other hand, can look at many pictures, though the overall acuity of each is not as good as for the modern biota. The perspective this provides is essential if we wish to take into account the effect of several cycles of biogeographic change on long-lived biotas. Since both the biogeographic approach and the paleobiogeographic approach have their strengths, both are essential. Without one or the other, we lack a complete picture of biogeographic patterns and the processes that govern these patterns. In short, biogeographic investigations should come to encompass paleobiogeography, for this "field of research extends over the whole Earth, not only as it now exists, but also during the continuous change it has undergone from the earliest geological epochs" (Wallace, 1857 in Brooks, 1984, p. 159).

Chapter 2

The Relevance of Hierarchy Theory to Biogeography and Paleobiogeography

2.1. Introduction

One of the strengths of paleobiogeography as a discipline is that it sits at the nexus or intersection of biological and geological research. Major conceptual advances in both of these subjects critically impact research in paleobiogeography. It is the association of the changing and coevolving spiral between life and geology, mediated by geography, that makes biogeography a very broad field (Brown and Lomolino, 1998). In this way it is similar to evolutionary biology, and Simpson's (1944) statement about that field also applies to biogeography: "the basic problems of evolution are so broad that they cannot hopefully be attacked from the point of view of a single scientific discipline" (Simpson, 1944, p. XV). I first want to demonstrate why this statement is true of evolutionary biology, and then I will go on to show why it is also true of biogeography.

In evolutionary biology, Simpson's statement partly reflects the fact that researchers can use different techniques to study evolution—e.g., molecular data, data from the fossil record, data concerning the structure of populations—and each of these techniques is generally utilized by a different type of evolutionary biologist. For example, molecular biologists, paleontologists, and population biologists, respectively, employ the three techniques listed above. However, the breadth of the field is not simply a matter of technique; or rather, the utility of these different techniques is indicative of something more basic: It reflects the hierarchical structure of nature.

That our universe in general and nature in particular is hierarchically structured on many fundamentally distinct levels is a topic that has been debated for some time, and many physicists have accepted it as a given (e.g., Feynman, 1965). In the biological sciences an incomplete list of hierarchically arrayed entities would include genes, ecosystems, organisms, species, organ systems, and higher taxa. We must look through this hierarchical lens to see biogeography as it truly is.

There is a long history of debate about the existence of hierarchies in biology. Some scientists have argued adamantly against the notion that nature is hierarchically structured, e.g., Dawkins (1976), although he later moderated his views (Dawkins, 1982). Similarly, Williams (1966) first argued against the existence of hierarchies in biology, and then also changed his opinion (Williams, 1992). These authors' early views on hierarchies are influential and are certainly accepted by some members of the scientific commnity. However, the arguments in favor of a hierarchical conception have, at least recently, perhaps been more extensive and as I see it more compelling [see Eldredge (1995) for discussion]. A long but not complete list of publications in which the authors argued for a hierarchical view of nature would include, though is not limited to, such early works as Smuts (1925), Dobzhansky (1937), and Mayr (1942), and later ones such as Eldredge and Cracraft (1980), Gould (1980, 1982, 1990), Hull (1980, 1988), Vrba (1980, 1983, 1989, 1993, 1996), Allen and Starr (1982), Arnold and Fristrup (1982), Eldredge (1982, 1985a, 1989a,b, 1995), Mayr (1982), Eldredge and Salthe (1984), Sober (1984), Vrba and Eldredge (1984), Damuth (1985), Salthe (1985), Brooks and Wiley (1986), Vrba and Gould (1986), Buss (1987), Brooks (1988), Brooks and McLennan (1991), Burns *et al.* (1991), Streidter and Northcutt (1991), Lieberman *et al.* (1993), Goodwin (1994), Rosen (1991), Lieberman (1995), Lieberman and Vrba (1995), and Valentine and May (1996).

Although it is clear from the extent of the foregoing list that there is an impressive array of references supporting the hierarchical viewpoint, I want to acknowledge that the other largely ahierarchical viewpoint is accepted in some quarters, and may have merit, especially when applied to certain types of problems. However, when we are trying to understand the mechanisms that have determined the complex structure of nature (the known biological universe) or the physical universe, simple reductionist claims are unlikely to meet with much success. I am not arguing that there is total acceptance by scientists and philosophers of science for the existence of hierarchies in biology. Still, the existence of such hierarchies would influence our views on which evolutionary patterns are worthy of study and would also determine the types of evolutionary processes that produce these patterns. Finally, it would have a bearing on what scientists believe to be the true entities out there in the world that we can study.

I would like my readers to treat the hierarchical viewpoint as one that may have some merit. Let us consider it and see where it gets us. In particular, for the purposes of this book, how does this putative hierarchical structure of nature influence our understanding of biogeographic patterns and processes?

I will present here a brief exposition of the hierarchical view as it applies to entities, patterns, and processes in biology, which will help frame further discussion of hierarchies in biogeography. The notion that nature is hierarchically structured is based on the claim that there are discrete entities that can be recognized in nature, and further that these entities are nested, with smaller ones inside of larger ones. Within each of these levels there is some autonomy; higher-level entities are not simple aggregates of lower-level entities (Eldredge, 1985a). Some have referred to this autonomy using the term emergent properties, which means that a species is more than just a collection of individual organisms. Something happens when a group of organisms forms a species and that species is not just the simple sum of its parts.

2.2. The Genealogical and Economic Hierarchies

There are two major varieties of hierarchically nested entities in the biological world. One includes genes, which are situated on chromosomes or housed within organelles, which are organized within cells (specifically, in most organisms, germ-line cells) that make up organisms, which in turn form parts of populations or demes of organisms, which comprise parts of species, which belong finally to higher taxa (for lists of this type see the extensive reference list on hierarchies cited above). A deme is a population of organisms of a given species joining together for reproduction or breeding (Wright, 1931). Note that breeding populations often form only during very specific times of the year, and often have very different structures and compositions than populations at other times.

In regard to what I mean by species, I recognize that the term may mean different things to different researchers. In fact, the scientific and philosophical debate about exactly what constitutes a species is very widespread and at times acrimonious. For the purposes of our discussion here, a variant of the biological species concept has some heuristic value. Specifically, a species is the largest cluster of organisms that recognize one another for the purposes of reproduction. Interested readers should see Vrba and Eldredge (1984), Eldredge (1985a,b, 1989a, 1993), de Queiroz and Donoghue (1988, 1990), Vrba (1989, 1995), and Lieberman (1992) as well as Chapter 7 in this volume for a discussion of species concepts. Finally, a clade is defined as an entity comprising an ancestor and all of its descendants.

All of the different genealogical entities mentioned above house some form of information (Eldredge and Salthe, 1984; Vrba and Eldredge, 1984; Eldredge, 1985a, 1986, 1989a). They are also more-making entities that reproduce and thus develop and modify the information, except for the higher taxa (Eldredge, 1985a) [these cannot reproduce without becoming paraphyletic and thus no longer clades (Vrba, 1996, pers. comm.)]. Thus, most of the members of the genealogical hierarchy are best seen as "reproducing 'packages' of genetic information" (Eldredge, 1986, p. 351). These entities belong to the genealogical hierarchy of Eldredge and Salthe (1984), Vrba and Eldredge (1984), Eldredge (1985*a*, 1986, 1989*a*, 1995), Brooks and Wiley (1986), Vrba (1989, 1995), Brooks and McLennan (1991), Lieberman (1995), and Lieberman and Vrba (1995). All are biological individuals in the sense of Ghiselin (1974) and Hull (1976, 1978, 1980). That is, entities such as individual genes, species, and higher taxa have histories, with birth and death points and some stability or constancy during their duration. In addition, they are held together by one of their characteristic features, the quality of reproduction or more-making, where more entities of like kind are created (Eldredge and Salthe, 1984; Vrba and Eldredge, 1984; Eldredge, 1985*a*, 1986, 1989*a*). For instance, a reproducing population or deme is held together by reproduction of its component organisms.

Although these entities, except for higher taxa, do make more individuals of like kind, there has been active debate about whether or not some of them may be what Dawkins (1976, 1982) and Hull (1980) termed replicators. A replicator, when it reproduces, gives rise to entities that are identical or nearly identical to itself. This is an important distinction, but not crucial for our discussion here.

There is also another major type of hierarchically nested entities. Involved in matter/energy transfer, exchange, and transformation, these entities include proteins nested within cells, inside of tissues, composing organs and thus organ systems, which comprise organisms, which belong to populations [or avatars in the terminology of Damuth (1985) and Eldredge (1986, 1989*a*)], which belong to ecosystems, which finally make up the biosphere. An avatar would comprise a population of a species that is interacting in some manner, perhaps cooperating to gather food or taking part in other social interaction not related specifically to reproduction. An ecosystem would be a set of populations or avatars of different species interacting in a given area with one another and with the abiotic environment.

These entities belong to the economic or ecological hierarchy of Eldredge and Salthe (1984), Vrba and Eldredge (1984), Eldredge (1985*a*, 1986, 1989*a*, 1995), Salthe (1985), Brooks and Wiley (1986), Vrba (1989, 1995), and Brooks and McLennan (1991). Different entities interact with one another within each level. That is to say, an economic population or avatar exists because of interactions between organisms that involve matter/energy transfer. In the list of entities in the economic hierarchy I use the term ecosystem rather than community. Altering this terminology would not appreciably change the conclusion of this discussion, which is to emphasize the fact that there is an overall distinction between the economic and genealogical hierarchies; however, it may lead to problems in logical consistency when I introduce the notion of geographic ranges as applied to entities in the economic hierarchy later on in this chapter, because it has been argued that communities may be defined strictly on the basis of the geographic range of a particular species [see Ricklefs (1979, p.669)]. Like the entities in the genealogical hierarchy, the entities in the economic hierarchy are individuals. They have histories, with birth and death points, and some stability or constancy during their duration, and they are held together by one of their characteristic features, the quality of interaction and matter/energy transfer.

The entities in both the genealogical and the economic hierarchies can be thought of as representing distinct individuals, rather than just aggregates of a lower level. This is because as I noted above a population is more than just a collection of organisms. It is something other than just the sum of its parts, and has what are termed emergent properties. For example, unique behaviors and interactions may arise when organisms come together to form a population, either a breeding population (a deme) or an economic population (an avatar). Such behaviors in the case of demes would include complex mating rituals, and in the case of avatars would include cooperating to obtain resources.

With this information in place it can readily be seen that because a hierarchically nested set of such genealogical and economic entities exists, the patterns that we see in these entities when we study them through time may be expected to be broadly hierarchical (Fig. 1). As is evident from Fig. 1 and the discussion above, most of the entities listed belong to only one of the two hierarchies. What is the significance of this? It influences our understanding of evolutionary and ecological patterns and processes in a fundamental way.

When we use the hierarchical perspective, evolution can best be viewed as the history of stability and change in the entities of the genealogical hierarchy. Then, the pattern of evolution is determined by the processes intrinsic to the

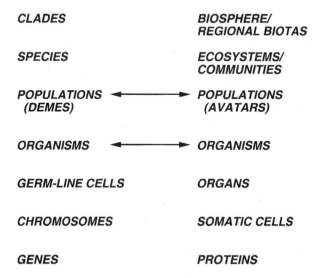

FIGURE 1. Hierarchically arrayed entities in both the genealogical (left-hand side) and economic (right-hand side) hierarchies. Arrows represent entities common to both.

different levels of that hierarchy as well as the interactions between the different entities comprising the economic and genealogical hierarchies (Eldredge, 1986). Within the two hierarchies there can be upward and downward pointing arrows or vectors of causation. The relevance and significance of this can best be seen by means of an example. We imagine that an organism dies. When this happens, its component germ cells and genes die, which would be downward causation. This death will also potentially influence the deme that the organism belonged to, as there will now be one individual less that can reproduce within it. To a much lesser extent it will influence the species the organism belonged to, as well as the higher taxa, e.g., monophyletic genera, families, within which that species is nested, which would be upward causation.

There may be evolutionary processes unique to each of these entities, such as genic selection, organismic natural selection, and species selection (Lieberman and Vrba, 1995). Further, analyses of real-world data suggest differences between patterns in the economic and the genealogical hierarchies. For example, Mayden (1988) conducted a detailed study of the freshwater fish of North America and found that there were major differences in the way in which communities were structured and organized and the patterns of phylogenetic relatedness among the taxa whose populations built up these communities [see also Brooks and McLennan (1991)].

Now, what about the entities in the economic hierarchy? These too would have arrows of upward and downward causation. If a population or avatar is eliminated it will affect the ecosystem of which it was apart, especially if it is a population of an important species within that ecosystem. It will also, to a lesser extent, affect the biosphere. This would involve upward causation. Similarly, if one thinks of an organism that dies, let us say owing to liver failure, it would also be a case of upward causation. However, because the organism died, the cells of the heart die too, which would involve downward causation. We can think of what goes on in the economic hierarchy as determining which players in the genealogical hierarchy survive and go on to repopulate the economic hierarchy. By analogy to the field of evolutionary biology, ecology would largely emphasize the analysis of entities in the economic hierarchy (Eldredge, 1985a, 1989a).

Although the fields of evolution and ecology largely emphasize different entities, they also share many commonalties. From the discussion above and Fig. 1, it is clear that there are a few entities common to both hierarchies. Note in particular that this is true for organisms and possibly populations. These are the entities which allow vectors of causation to extend across the hierarchies (Eldredge and Salthe, 1984; Vrba and Eldredge, 1984; Eldredge, 1985a, 1986, 1989a), and interdigitation between the fields of evolution and ecology. In this case, an analysis of the entities that appear in both hierarchies, such as organisms and populations (demes and avatars, when they coincide), would have immediate relevance for researchers in both evolutionary biology and ecology. What happens to entities in the genealogical hierarchy can potentially influence the economic hierarchy and vice versa, but these effects would be filtered through different levels in each of the two hierarchies and then impact across those entities that span them both. To see how this works, let us go back to the death of an organism, which, as I noted, potentially influences several representatives of entities within the genealogical hierarchy via upward and downward causation from the organism level. There can also be vectors of both downward and upward causation within the economic hierarchy extending from the same level.

2.3. Hierarchies and Evolution

The foregoing hierarchical formulation has important implications for our understanding of evolutionary theory. For example, since the time of Darwin (1859), natural selection has been seen to be one of the most important mechanisms mediating evolutionary change. Hierarchy theory fits in well with this conception when we start to look at how selection processes operating at different hierarchical levels may work [see, e.g., Hull (1980), Sober (1984), Vrba (1984a, 1989, 1995), Eldredge (1989a), Lieberman et al. (1993), Lloyd and Gould (1993), Lieberman (1995), Lieberman and Vrba (1995)], and at the distinction between sorting and selection [see Vrba and Gould (1986), Vrba (1989), and Lieberman and Vrba (1995)]. For selection to occur, there has to be reproduction. [Some scientists, such as Dawkins (1976, 1982) and Hull (1980) have argued that replication as well as reproduction, is necessary. These reproducing entities would obviously be representatives of the genealogical hierarchy. Anything that cannot reproduce cannot be selected. As I noted above, clades cannot reproduce without becoming paraphyletic and therefore no longer clades, so they cannot be selected, counter the claims of Williams (1992). Further, to have selection there must be interactors or vehicles in the parlance of Hull (1980) and Dawkins (1976, 1982), respectively. Such interactors would be representatives of the economic hierarchy. Thus, for selection to occur, both replication and interaction would be required. The only entities that could be selected would belong to both the economic and the genealogical hierarchies (Hull, 1980; Eldredge, 1985a, 1989a; Vrba and Gould, 1986; Vrba, 1989). Organisms, populations, and certain cell types may belong to both hierarchies, but any entity that does not, cannot be selected.

The formulation of the hierarchical viewpoint in biology has also demonstrated the impossibility of certain concepts in evolutionary biology. For example, the notion that there is such a phenomenon as community evolution. When the communities of any given region are studied through time they appear to change. For example, Van Valkenburg (1988), and Webb (1989) have discussed the way in which the terrestrial mammalian communities of North America have changed over millions of years, but communities cannot evolve as populations or species do because they cannot reproduce or make more of themselves.

Other ideas or concepts in evolutionary biology that have been illuminated by this hierarchical framework include whether or not higher taxa can be held to sit on adaptive peaks by occupying a particular niche [see Eldredge (1985a)]. Although such a viewpoint was prevalent in some very important books in evolutionary biology, such as those by Dobzhansky (1951) and Simpson (1944), it appears that it is no longer valid. For example, the family Felidae cannot be said to occupy a particular place in the economic hierarchy. Populations of different species of cats are carnivores in different local ecosystems, but they do not interact with the rest of the world's animals and plants as a single economic entity (Eldredge, 1986, 1989a; Vrba, 1995). By the same token, a species cannot be seen to be a member of the economic hierarchy unless it is reduced to a single population that functions as both an avatar and a deme, because a species does not interact as a whole with any other species. Rather, different populations of a species may interact with populations of other species. Therefore the boundaries of a species will generally not be boundaries between economic entities (Ricklefs, 1979; Eldredge, 1989a; Vrba, 1995). For instance, the mountain lion is found from northern North America to the southern tip of South America. In different regions it feeds on different prey and lives in different types of habitats, from high mountains to deserts to swamps. Its range cannot be said to be equivalent to that of a single ecosystem. Rather, different populations of mountain lions play the role of top carnivore, or one of the top carnivores, in very different ecosystems.

Probably one of the most important conceptual developments engendered by the hierarchical expansion is the status of theories that posit that evolutionary processes operating at a lower hierarchical level can be smoothly extrapolated to explain patterns in a higher-level genealogical entity. Specifically, some theories about the evolutionary process are based on the premise that we can extrapolate processes at one hierarchical level, let us say the genetics of populations, to explain the diversification of a clade. The problems with this extrapolationist approach in evolutionary biology have been discussed extensively by Gould (1980, 1982, 1990). If each of the levels of the genealogical hierarchy, such as genes, organisms, populations, and species, represents a distinct, real entity, then processes that operate at one level may not translate freely to a lower or higher level. Each level is distinct, with its own characteristic emergent properties, and, potentially, distinct processes.

This notion can be best explained by means of an example. Processes that influence population genetic structure can play a role in structuring evolutionary patterns among species, but they are not the sole determinants of these patterns. The existence of a species as a distinct hierarchical level with its own emergent properties means that it is more than just an aggregate of populations, and therefore processes within each of those populations do not simply translate into changes within species. We can take as a case in point the phenomenon of stasis within species. It appears that most species throughout much of their existence are stable entities that change relatively little morphologically over millions of years (Eldredge and Gould, 1972; Gould and Eldredge, 1977), rather than changing gradually over a long, protracted interval. This has been documented by many studies including those of Stanley and Yang (1987), Geary (1990), Jackson and Cheetham (1990), Lich (1990), Cheetham *et al.* (1994), Lieberman *et al.* (1994, 1995). [Some other studies do suggest alternative patterns to obdurate stasis, and readers can refer to Geary (1990), and Sheldon (1987) for possible counterexamples.]

There are several reasons why we might expect species to show an overall pattern of stasis, one of them being their simple structure (Lieberman *et al.*, 1994, 1995; Eldredge, 1995; Lieberman and Dudgeon, 1996); they are made up of populations of reproducing organisms or demes. A single deme of a species might change significantly through time, but this does not require that the mean morphology of the entire species change. Species are typically made up of several demes, and the total change in the morphology of a species is equal to the sum of the changes in all of its constituent distinct demes. Each deme is likely to undergo a separate adaptive history and thus display morphological changes that are different and independent from those of other conspecific demes. The net result is that these changes will generally cancel one another out, resulting in no net change in the morphology of any individual species (Eldredge, 1989*a*, 1995; Lieberman *et al.*, 1994, 1995; Lieberman and Dudgeon, 1996).

The analysis of changes in the morphology of demes through time is clearly worthwhile; it just does not necessarily translate with smooth extrapolation to any statement about what happens to the morphology of any (multidemic) species through time. Again, this is due to the fact that the biological realm, and the entities that evolve, are arrayed hierarchically.

The fallibility of extrapolationism, demonstrated for evolutionary biology, also applies to ecology. The different entities in the economic hierarchy are more than simple sums of the parts at the immediately underlying level; i.e., they have emergent properties (Allen and Starr, 1982; Eldredge and Salthe, 1984; Salthe, 1985; Eldredge, 1986).

Hopefully, this discussion has provided some broad outlines in regard to the types of hierarchies that exist in nature and also some insight as to how this hierarchical formulation can influence the way we view evolution and ecology. I am going to argue that in addition to the implications that I have outlined thus far this hierarchical formulation impacts critically on the fields of biogeography and paleobiogeography.

2.4. Hierarchies and Biogeography

Hierarchies in biogeography can encompass several features, including geographic ranges. For example, the range of an Order of birds is greater than the geographic range of a Family contained within that Order. This will almost always be true except in trivial cases such as monotypic higher taxa. In fact, Wallace (1855) and Darwin (1859) took this hierarchical arrangement of geographic ranges as prima facie evidence that life had evolved, because it reflects a pattern of differentiation via common descent [see, e.g., Darwin (1872, p. 344)].* Both Wallace and Darwin recognized that if life has evolved, there should be a nested pattern of descent, with a higher taxonomic level reflecting an earlier origination in time and necessitating the occupation of a broader geographic range.

To address other critical issues in regard to the way hierarchies relate to biogeography, it is perhaps easiest to begin by discussing the different types of biogeographical entities and how they are arrayed hierarchically. In the earlier discussion of hierarchical entities in biology, they constituted a discrete set in the natural world. As biogeography concerns both life and geography, if hierarchically arrayed biological entities do exist that fact is clearly relevant (Brooks, 1988; Brooks and McLennan, 1991).

There is a potential for recovering biogeographic patterns within both the genealogical and economic hierarchies. Patterns within the former will involve entities that reproduce and preserve information. These will be governed to some extent by evolutionary processes and they will also reflect historical events. Patterns within the latter will involve entities that undergo matter/energy exchange and transfer. These will be governed by ecological processes, though again they will also reflect historical events.

It is the existence of these two hierarchies that should delineate the boundaries between the two subdisciplines of biogeography—historical biogeography and ecological biogeography. However, as they have been traditionally defined the first focuses on the role of geological and climatic factors and the second on that of ecological factors in biogeography. There is also interdigitation between these two subdisciplines as some entities are shared between the two hierarchies—specifically, organisms, and potentially populations (to the extent that avatars and demes overlap)—and this opens the way for a potential unification of these various subdisciplines and approaches.

2.4.1. Biogeographic Patterns in the Genealogical Hierarchy

If we refer back to the list in the Preface we note that some of the questions that historical biogeographers have posed over the last few centuries include questions 1–4 and 7–9. A fundamental concern of historical biogeographers is to determine whether or not taxa show similar patterns of evolutionary differentiation across geographic space, i.e., among several different groups is the taxon in one area (the hatched elliptical area) always the closest relative of the taxon in another area (the circular area) (Fig. 2). There are biogeographic patterns among entities at several levels of the genealogical hierarchy, e.g., patterns of organisms, populations or demes, species, and monophyletic taxa. Since each of these levels has its own unique emergent properties, and possibly also its own intrinsic evolutionary processes, patterns at a lower level cannot be smoothly extrapolated to a higher one. Further, patterns at each of

---

*All citations to *On the Origin of Species by means of Natural Selection* will be to the 6th edition, unless otherwise noted, as this is the edition most readily available.

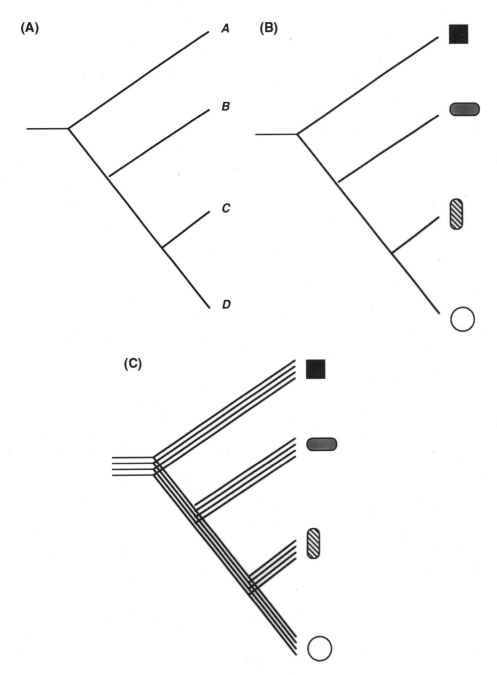

FIGURE 2. (A) A cladogram relating four hypothetical species. (B) The geographic occurrence of these species has been substituted for the taxon name; the different shapes represent different geographic regions. (C) Four separate clades showing the same pattern of biogeographic relationship among areas, evidence for congruent patterns of evolution across geographic space.

these levels should operate on an approximate timescale that is also hierarchical, with the time interval we associate with biogeographic differentiation of higher taxa longer than the one for populations. Brooks (1988) and Brooks and McLennan (1991) were the first to recognize this phenomenon.

When ecologists and evolutionary biologists consider various spatial and temporal scales, they find that very different types of evolutionary and ecological patterns and processes can occur (Brooks and Wiley, 1986; Ricklefs, 1987; Brooks, 1988; Brown and Maurer, 1989; Brooks and McLennan, 1991). For example, the types of evolutionary, geological, and/or climatic processes that influence biogeographic patterns within species might differ from those that influence patterns among species or within clades. Some of the processes that influence clades include allopatric speciation, vicariance, and dispersal. The time frame associated with biogeographic patterns within clades would have to be at least as long, and probably longer, than the one we associate with intraspecific biogeographic patterns.

2.4.2. Biogeographic Patterns in the Economic Hierarchy

Just as there are historical biogeographic patterns in each of the entities in the genealogical hierarchy, there are potentially ecological biogeographic patterns in those of the entities in the economic hierarchy. Such patterns might involve the way in which ecosystem differentiation is related to geographic distribution, which was studied by Brown and Maurer (1989), who considered how physical space and nutritional resources were divided in the diverse extant North American avian and mammalian clades. Ecological biogeographers might also focus on how the geographic distribution of a population is influenced by the distribution of microhabitats and the vagility of their component organisms (Brooks and McLennan, 1991).

Thus, it is clear that within each of these hierarchies there may be very distinct types of biogeographic patterns that are arrayed hierarchically. Patterns and processes at a lower level cannot necessarily be extrapolated to higher levels and vice versa. An important issue then becomes how we relate the biogeographic patterns in the entities of these different hierarchies.

2.4.3. Relating Biogeographic Patterns across the Hierarchies

It has been argued that as the spatial and temporal scale of any biogeographic study increases, the influence of phylogenetic or historical biogeographic signal is more strongly felt (Brooks and McLennan, 1991). That is, the extent to which the larger-scale entities in the economic hierarchy show some biogeographic signal is overpowered by the evidence for biogeographic signal and differentiation in the larger-scale entities in the genealogical hierarchy. Brooks *et al.* (1981), Ross (1972, 1986), and Mayden (1988) demonstrated that when we look at large assemblies of taxa we often find that the way these taxa are related is a better predictor of their geographic distribution than their ecological characteristics. However, in other cases, at broad spatial and temporal scales an ecological biogeographic signal may be present and strong (Brown and Maurer, 1989; Brown and Lomolino, 1998). (The argument about the relative prevalence of signals for both ecological and historical biogeographic patterns, is one of the oldest debates in the field, and I will explore its history more fully in the next section.)

Because the economic and genealogical hierarchies share one—and possibly two entities—biogeographic patterns can coincide across them. For example, when we approach topics such as biogeographic differentiation within populations, there may be close congruence between historical and ecological biogeographic patterns. In other words, the geographic distribution of organisms within populations related evolutionarily via a pattern of descent and the geographic occurrence of organisms interacting with one another in populations might be congruent. However, as we climb each of the hierarchies, we move further away from these bridging entities and so biogeographic patterns become more dissimilar. Ecological biogeographic topics focusing on higher-level entities might include a study of why it is that throughout the biosphere taxonomic diversity correlates with latitude. This topic also takes in genealogical entities, as there are similarly shaped latitudinal diversity gradients within individual higher taxa, but it relates primarily to ecology and entities in the economic hierarchy (Stevens, 1992). Specifically, why can we pack more avatars into a tropical ecosystem than into a boreal one? By contrast, historical biogeographic questions might focus on whether or not a set of disparate taxa show similar patterns of speciation associated with the establishment of some geographic barriers by a series of geological events. Within the regions being studied, ecosystems will of course be present, but these entities may not be of interest to the historical biogeographer and thus will not be sampled, while the species that provide the avatars that make up these ecosystems will be sampled.

2.5. Climate Change and Biogeographic Patterns

Other differences between historical and ecological biogeographic patterns can be elucidated by considering how global change—encompassing climatic and geologic—affects biogeographic patterns in the different entities in the two hierarchies. Some short-term climatic changes that are governed by distinct astronomical cycles influence biogeographic patterns (Huntley and Webb, 1989). These include daily, seasonal, and annual cycles, which in turn influence the behavior of organisms and can lead to large-scale migrations. For example, herds of wildebeest stream majestically across the African plains in response to changes in solar radiation and their influence on temperature and humidity; terns yearly span the globe; caribou undertake massive and awe-inspiring migrations. Further, these large annual or seasonal movements are often associated with reproduction. That is, yearly climate changes owing to seasonal cycles act as cues for animals to undertake major migrations. Gray whales travel from waters in the North Pacific brimming with marine zooplankton to quiet waters off of Mexico's Baja Peninsula to calve. Warblers and other songbirds fly from the teeming, diverse Central and South American tropics to the United States and Canada to find mates and nest. There are also annual and decadal fluctuations in sea-surface temperatures, reflecting oceanographic changes, that alter the ranges of species and communities (McGowan *et al.*, 1998).

There are even longer-term cycles operating over millennia. Such cycles were documented by Hays *et al.* (1976), Berger (1980), Imbrie and Imbrie (1980), and others. Most of the major peaks in these cycles are given below in terms of their temporal duration, from shortest to longest, and are referred to as: (1) precession, the wobble of the Earth's axis, which at present cycles approximately on the order of 19,000 and 23,000 years; (2) obliquity, the tilt of the Earth's axis, with a cycle of approximately 41,000 years; and (3) eccentricity, the distance between the Earth and the Sun, with a cyclic periodicity of approximately 100,000 years. These cycles can profoundly influence global climate. For example, the profound patterns of glacial advance and retreat (referred to as glacial and interglacial periods) on Earth during the Quaternary period were first thoroughly documented by Agassiz (1840), and since then by hundreds of other authors. These cycles were driven by subtle changes in insolation resulting from astronomical cycles and amplified by other cascading effects of the overall climate system of the Earth. Depending on the position of the continents, oceanic current systems, and mountain belts, climate change associated with this Milankovitch cyclicity may or may not be particularly profound.

Now let us consider how these longer cycles influence the distribution of entities, i.e., biogeographic patterns, in the genealogical and economic hierarchies. First, for the genealogical hierarchy: Obviously, the Milankovitch climate cycles operate over timescales much longer than the lifetime of any organism and probably almost all populations (and of course this is also true of other even smaller entities). Instead, they would most profoundly influence biogeographic patterns within species and also potentially within clades. Interestingly, the changes in temperature caused by these Milankovitch cycles may be of a much lower magnitude than the changes in temperature between an average day and night, but they can still have a profound impact on global mean temperature change (Huntley and Webb, 1989). The patterns of climatic change associated with Milankovitch cycles have been particularly well documented for the Quaternary period.

It turns out that these cyclical patterns of climate change also influence the distribution of organisms. Between glacial and interglacial periods species can come to have dramatically different distributions, all in an effort to track the environment that they prefer. For example, Coope (1990) documented how species of beetles, now restricted to the Mediterranean Basin, occurred as far north as the British Isles during warmer interglacial periods, and this type of shift in the range of insect species reflects a general pattern throughout Eurasia (Coope, 1979). A number of tree species, exemplified by taxa of the oak genus

Quercus, have seen their range expand and contract during the various glacial and interglacial periods (Davis, 1986; Huntley and Webb, 1989; Foster *et al.*, 1990). Mammal species show this as well (Graham, 1986, 1992; Foster *et al.*, 1990; Graham *et al.*, 1996), as do fossil marine macroinvertebrate species, as elucidated by Vermeij (1978), and marine microfossil species, as extensively documented by Bennett (1990).

Sometimes species tend to change little if at all morphologically throughout these episodes of long-term climate change; they simply track their preferred environment (Huntley and Webb, 1989; Bennett, 1990; Vrba, 1995, 1996). However, according to Elizabeth Vrba (1980, 1983, 1995, 1996) these types of climate changes may also produce episodes of speciation and extinction, influencing the diversity of clades and thus potentially their ranges. This falls under the rubric of what she has termed the Turnover-Pulse hypothesis. As the climate changes and species try to track their preferred environments, populations can become isolated, which tends to encourage evolutionary change. When the preferred habitat of a species shrinks too much owing to climate change, such species can go extinct. These patterns of speciation and extinction will be expected to be replicated across several clades. Further, species and clades tend to have cohesive responses to these types of climatic changes, i.e., species and clades can be recognized before, during, and after these changes.

Most of the studies on the biogeographic responses of clades and species to Milankovitch-style climate cycles have focused on the fossil record of the late Cenozoic, considering primarily the last 2–3 million years, mainly because our limits of resolution in the fossil record are much better for these periods and, further, owing to a large body of rsearch (e.g., deMenocal, 1995) we have an excellent understanding of how climate changed during this interval. Perhaps the one significant counterexample to the plethora of late Cenozoic and Quaternary analyses are the studies by Paul Olsen and others on the Mesozoic freshwater lakes of what is now eastern North America. These lakes, similar to the modern-day East African rift system of lakes, formed as the North American continent separated from Europe and Africa roughly 200 million years ago and were abundantly populated by fish. A large amount of statistical, paleontological, and sedimentological evidence gathered by Olsen (e.g., Olsen, 1984, 1986) has demonstrated that these lakes expanded and contracted during Milankovitch-scale climate cycles. Further, the expansions and contractions concomitantly influenced the distribution of fish species and thus fish communities (Olsen, 1984, 1986; McCune, 1987).

To finish up this part of my discussion, in terms of biogeographic patterns in the entities in the genealogical hierarchy, we can see that they move around in response to astronomical cycles that drive daily, seasonal, and longer-term climate cycles. Now what about entities in the economic hierarchy? Obviously organisms are moving around, but what about higher-level entities such as communities or ecosystems? These have also been studied extensively. It turns out that the populations of the different species that make up communities and ecosystems tend to migrate individualistically in response to the Milankovitch-cycle-driven environmental changes (Davis, 1976; Huntley and Webb, 1989; Bennett, 1990; Foster *et al.*, 1990; Graham, 1992; Vrba, 1993, 1996; and references therein). This is because ecosystems appear to represent only avatars of different species that occur together and interact with one another because of their shared tolerance for several environmental parameters (Vrba, 1993). Then, the region that they occur in represents a place where there is a concatenation of all of those environmental parameters.

Climate changes cause both major and minor shifts in a variety of environmental parameters. As the climate changes, avatars of different species will track their preferred environmental parameters, often over very long distances. After a substantial climate change, the exact combination of environmental parameters needed to sustain all the important avatars of a given ecosystem are not likely to be met, so different avatars will move in different directions to track their preferred environments. The results of these large-scale movements is that avatars of species that were in contact tens of thousands of years ago are no longer, and ecological communities and ecosystems appear to have been massively shuffled through time. Thus, over long timescales, communities and ecosystems are ephemeral historical entities (Huntley and Webb, 1989; Bennett, 1990; Foster *et al.*, 1990; Graham, 1992; Vrba, 1993, 1996). Over the roughly 20,000- to 100,000-year intervals of Milankovitch cycles, we come to see how, at least for the purposes of biogeographic theory, the entities in the genealogical hierarchy differ fundamentally from those in the economic hierarchy. Communities and ecosystems are evanescent in the face of these cycles whereas genealogical entities such as species and clades can persist and indeed are even obdurately stable.

This phenomenon makes it generally more difficult to study biogeographic patterns in an entity such as an individual community or ecosystem than in a genealogical entity such as a clade. An example here may help. We can think of an individual community that occupied a small stretch of Atlantic beach in eastern Georgia a few tens of thousands of years ago, and then imagine that there were climate changes associated with Milankovitch cyclicity. This community will be expected to have no real persistence in the face of such climatic changes, because different populations of species will migrate in different directions, tracking their preferred environmental parameters, and in effect breaking up the interactions that brought that community into being. It will be hard to study biogeographic patterns within this community during climatic change because something that quickly ceases to exist cannot show any prolonged pattern. What a scientist is likely to witness is first community disassembly and then an entirely new community springing up in the region being studied.

However, although communities may not persist through time, an ecological biogeographer might look for general rules or ecological processes associated with their establishment that relate in some way to the geography of the region that they occupied. For instance, maybe the most diverse communities are always found in a subtidal habitat whereas in a supratidal habitat, only a few, physiologically tolerant species are found, regardless of the climatic regime at any one time and irrespective of the species that provide the avatars that make up the community. The pattern being searched for relies on data from several different historical individuals over several different time periods.

By contrast, the approach in historical biogeography is different. An historical biogeographer might study several individual species believed to have been strongly affected by one or more periods of climate change. We might consider, e.g., species of marine invertebrates such as clams, snails, and crustaceans distributed throughout the present-day eastern seaboard of the United States, including that small stretch of beach in eastern Georgia just alluded to. A phylogenetic biogeographer might look at patterns of molecular genetic differentiation within each of these species to see how the different populations of any given species are evolutionarily related to one another (Avise, 1986; Avise *et al.*, 1987). The type of pattern a phylogenetic biogeographer might find shows that the populations of the species of clams, bivalves, and crustaceans that are to be found in eastern Georgia are most closely related to populations of the same species in South Carolina rather than to those in Delaware. Further, there may be a reason to believe that during times of glaciation, when sea level fell significantly, a spit of land might have emerged that acted as a fairly long-lived terrestrial barrier separating the marine invertebrate populations in South Carolina and Georgia from their conspecific populations in Delaware. Evolutionary processes that might be studied in this system include the role of geographic barriers in facilitating evolutionary change within species, with the pattern being examined involving repeated similarities in historical, genealogical entities or individuals during one period of time.

Significantly, when a pattern such as this is discerned in several populations it means that there is a potential for continuity of ecological interactions and associations (Avise, 1986, 1992; Kluge, 1988; Zink, 1991). Although there are many cases in which historical and ecological biogeographic patterns diverge, when this type of pattern is uncovered it may mean that they coincide, and it can be viewed best as the phenomenon of coevolution at the level of populations (see Brooks, 1985; Brooks and McLennan, 1991). Thus, not only have entities in the genealogical hierarchy differentiated along lines of geographic space, but entities from the economic hierarchy have done so as well. This is prima facie evidence for both phylogenetic and ecological biogeographic differentiation, which can occur because populations (as demes and avatars, respectively) can be entities in both the genealogical and the economic hierarchy.

2.6. Geological Change and Biogeographic Patterns over Even Longer Timescales

There are characteristic processes or cycles that operate on even longer timescales than Milankovitch cycles that can structure biogeographic patterns in the entities of the genealogical and economic hierarchies. One prominent set of processes relates to Earth history events, especially those mediated by plate tectonics. Plate tectonics is discussed in greater detail in Chapter 6, but the particular phenomena of special relevance here are continental fragmentation or rifting and continental collision. Plate tectonics on Earth reflects the fact that the globe is broken up into a series of more than 20 plates that are slowly but inexorably moving. New plate material is created at zones of upwelling such as the Mid-Atlantic Ridge, and material is returned to the Earth's interior at oceanic trenches. Further, over the course of time, supercontinents, large groupings of plates, are assembled and then split apart.

Plate tectonics drives a series of events that occur over timescales of millions to tens of millions, and maybe even hundreds of millions, of years, which can have a profound influence on biogeographic patterns. For example, continental rifting can cause populations of species to be isolated, thereby facilitating speciation via what is termed allopatric speciation and vicariance (Croizat *et al.*, 1974; Nelson, 1978; Rosen, 1978, 1979); Platnick and Nelson, 1978; Brooks *et al.*, 1981; Brooks, 1985, 1988; Wiley and Mayden, 1985; Mayden, 1988; Wiley, 1988a,b; Brooks and McLennan, 1991; Lieberman and Eldredge, 1996).

Vicariance is discussed more fully in Chapters 5 and 6, but a classic example would involve a single terrestrial species, let us call it species A, that is distributed across a large land mass (Fig. 3). We imagine that this land mass splits apart roughly down the middle owing to some rifting event and that an ocean fills the intervening space between what has become two continents. Now, populations of the original species on the two separate continental fragments or cratons will be isolated, separated by an impenetrable barrier. As these populations become isolated they tend to differentiate, and eventually new species will be formed; let us call them B and C. Each of the new species will inhibit a different, narrower range than the original ancestral species. Now that these cratonic plates are separated, they, along with their component organisms, including species B and C, will undergo driftlike motions. Over very long periods of time this drift can lead to dramatic changes in the climate they experience, especially when they move from low to high latitudes or vice versa. From the previous discussion of how climate change can drive evolution and biogeographic changes in species and clades, we can see how this might influence biogeographic patterns associated with the evolution and geographic distribution of the species and clades that occupy these plates.

Of course plate tectonic events do not only separate faunas, inducing biogeographic differentiation. Tectonically mediated events such as continental collision can also bring formerly separated faunas into contact with one another producing a pattern of congruent range expansion in several groups. This has been termed geodispersal (Lieberman and Eldredge, 1996) and is discussed more fully in Chapter 6. A classic example of the initial phase of geodispersal would involve two terrestrial species, A and B, each distributed across a single land mass and thus separated from one another by an intervening ocean (Fig. 4). We imagine that these two continents are moving

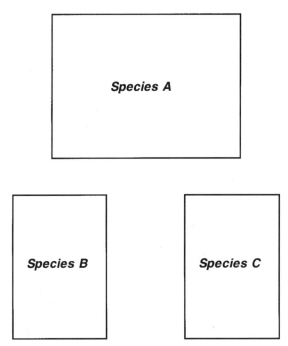

FIGURE 3. An example of vicariance, where a single ancestral species differentiates into two descendant species after the formation of a geographic barrier within its ancestral range: (Top) Terrestrial species A distributed across a continent; (Bottom) a rifting event bifurcating the continent produces two new, smaller areas; eventually two new species, B and C, are found in the new areas.

slowly toward one another and then collide so that they become a single large continent. When they collide, at least initially, each species may expand its range into the newly available space.

Eventually, continental collision can lead to the elevation of a mountain range, such as the Himalayan range, which is the rsult of a collision between India and Asia. It is conceivable that such a mountain range could, in time, become a barrier to the free movement of many terrestrial organisms, eventually isolating populations of species on either side of the range, which would lead to vicariant speciation. Thus we can see how geodispersal and vicariance, mediated by plate tectonic events, might operate hand-in-hand to influence biogeographic patterns. Plate tectonic events tend to impact on several different groups of species in the affected area. Many analyses have shown that there is an important relationship between times of major plate tectonic change and major evolutionary change (Hallam, 1967; Valentine and Moores, 1970, 1972; Valentine et al., 1978; Brooks and McLennan, 1991; Dalziel, 1997; Lieberman, 1997). Since there is also an association between plate tectonics and biogeographic patterns, there must be some correlation between major evolutionary change and major biogeographic change. (The

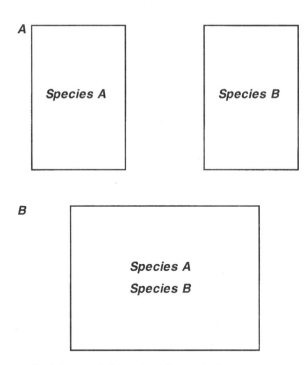

FIGURE 4. An example of the initial phase of geodispersal when two species occur in two separate areas that are separated by a geographic barrier. When, the barrier between the two areas falls, each species can expand its range into the other area: (Top) two terrestrial species, A and B, on continents separated by an intervening ocean; (Bottom) these continents collide and each of the species can expand its range.

close coupling between evolutionary theory and biogeography will be considered in greater detail in several subsequent chapters.)

Now, what about biogeographic patterns in entities of the economic hierarchy over very long timescales? Since large-scale economic entities such as communities and ecosystems may be largely ephemeral over timescales of tens or hundreds of thousands of years, one might expect that these entities, and other, even larger, ones such as regional biotas, would be ephemeral over even longer timescales. Most evidence also suggests that communities and ecosystems did not have cohesive biogeographic patterns during Milankovitch cycles. Then, if we follow this logic, communities, ecosystems, and regional biotas over the even longer timescales associated with plate tectonic events should also not have cohesive biogeographic patterns. However, this has not yet been demonstrated conclusively. In fact, it is a topic of active debate among paleoecologists and paleobiogeographers.

Some scientists have argued, particularly Jackson (1992), Morris *et al.* (1995), Jackson *et al.* (1996), and Pandolfi (1996), that although communities and ecosystems appear to coalesce and break up over the timescales of Milankovitch cycles, over even longer timescales, and across broad geographic regions, there are in fact patterns of stability and persistence in communities, ecosystems, and regional biotas. Brett and Baird (1995) have referred to these hypothesized patterns of long-term stability in large-scale entities of the economic hierarchy as coordinated stasis, a pattern first recognized in fossiliferous rocks from the Appalachian Basin of eastern North America. These rocks record a time when much of North America was a shallow seaway, populated by invertebrate groups such as rugose corals, brachiopods, and trilobites. Studying the rocks of this region, they documented the existence of faunal packages that represent 5 to 10-million-year blocks of time in the fossil record, which contained a large number of essentially unique species. The fauna in any given block shared very few species with either the overlying or the underlying faunal package, so faunal boundaries appeared to represent extinction events, as well as origination and/or immigration events.

Brett and Baird (1995) argued that during the existence of a fauna, the species within the fauna, as well as the communities built up from the populations of those species, tended to change little if at all. They suggested that the presence of each fauna was preceded by a wholesale extinction, due perhaps to some major environmental change. Then, in a very short time, at least geologically, new species invaded the region, with some speciation events occurring as well, leading to the establishment of a new regional fauna. Once the new fauna was established, few new species appeared, and its composition, as well as the communities within it, remained essentially unchanged.

Building on this work, Morris *et al.* (1995) put forward a mechanism to explain the pattern of coordinated stasis that they called ecological locking. Their mechanism involved a potentially new set of higher-level ecological processes that relate to a fauna's ability to withstand invasion by taxa from outside its region and to resist environmental perturbations. The hypothesis of coordinated stasis and the mechanism put forward to explain it if they are correct, would be very relevant to biogeography and paleobiogeography because they suggest the possibility that higher-level ecological entities have a characteristic position in space, show stability over time, and owe their maintenance, at least in part, to ecological mechanisms. Such higher-level ecological entities, if coordinated stasis holds true, would be regional biotas that occupy some broad geographic area, such as the Appalachian Basin, and show long-term biogeographic integrity and stability. Paleobiogeographic patterns that might be studied in these faunas include determining whether there was an association between tectonic events and patterns of faunal change and how new faunas assembled biogeographically.

Although the hypothesis of coordinated stasis, if true, would have important implications for the status of ecological paleobiogeography as a field, it has not yet met with universal acceptance. Several authors, including Buzas and Culver (1994), Bennington and Bambach (1996), Holland (1996), Holterhoff (1996), Westrop (1996), and Patzkowsky and Holland (1997), have argued that coordinated stasis is not a universal pattern in the fossil record, concluding that over long timescales and across broad spatial scales, entities such as communities, ecosystems, and regional biotas are not cohesive and stable. Although the results of these studies are open to interpretation, denying coordinated stasis as a general hypothesis would refute the concept of biogeographic patterns in higher-level entities of the economic hierarchy or, at least, the existence of patterns that would be analogous to those in comparable levels of the genealogical hierarchy. This would imply a significant disjunction between our ability to study biogeographic patterns in the two hierarchies at higher levels. In particular, individual clades would persist through long term geological changes such that the effects of geographic changes on individual, historical entities could be monitored. However, individual communities, ecosystems, and regional biotas would break apart in the face of such profound geological changes, making it impossible to monitor biogeographic changes within any one of these types of ecological entities over long timescales.

If coordinated stasis is corroborated by future studies, then entities such as communities, ecosystems, and regional biotas will tend to show some higher-order stability in the face of tectonic changes, such that biogeographic patterns within these individual ecological entities can be monitored over long periods of time. Then, individual entities from both hierarchies could be studied using partially congruent approaches. Further, if ecological locking is the operative mechanism, then stability in regional biotas would be governed by ecological processes. This would be an important validation of a research program in ecological paleobiogeography that would focus on patterns of stability and change in higher-level ecological entities and the mechanisms that govern these patterns.

One point that does need to be emphasized in the discussion of the tests of the coordinated stasis hypothesis conducted thus far is that many of the studies that have demonstrated the evanescence of communities and ecosystems during timescales roughly equivalent to Milankovitch cycles did not consider larger-scale entities in the economic hierarchy such as regional biotas. These, unlike communities and ecosystems, would be more likely to show the coordinated stasis pattern. Whatever the outcome of this debate, it is likely that this will be an active and exciting area of research engaging paleoecologists and paleobiogeographers for many years to come.

2.7. Mass Extinctions and Biogeography

There may be one additional cycle that operates on long timescales and influences biogeographic patterns in entities of both the genealogical and the economic hierarchy. Raup and Sepkoski (1986) and Raup (1986) have argued that there is a significant periodicity to patterns of extinction in the fossil record. Specifically, they claim that roughly every 26 million years over at least the last 260 million years there has been a major episode of mass extinction that eliminated as many as 95% of all the species on the planet. Further, the apparent regularity of these extinction events led them to invoke an astronomical forcing mechanism, specifically, asteroid and/or comet showers driven by gravitationally mediated perturbations caused by a possible brown dwarf or "Nemesis" star orbiting our solar system (Raup, 1986).

Patterson and Smith (1987) raised legitimate concerns about the database from which the evidence for extinction events was drawn, which may cast the entire periodic pattern into doubt. However, even if the periodicity of the pattern is open to question, it is clear that mass extinction is a real phenomenon (Stanley, 1987; Eldredge, 1991). Mass extinctions can obviously influence biogeographic patterns in both hierarchies to a very great extent because they cull a large proportion of the standing diversity of species and clades present at any one time, thereby also eliminating or fundamentally altering many of the Earth's communities, ecosystems, and regional biotas. The entire biosphere was surely affected by the events at the end of the Permian when possibly as many as 96% (Raup and Sepkoski, 1986; Eldredge, 1991) of all species went extinct [but see Stanley and Yang (1994) for a revised view of the Permo–Triassic extinction].

Mass extinction events should be associated with biogeographic patterns, and some authors have considered this point in great detail. Specifically, Jablonski (1991) and Smith and Jeffery (1998) have studied the mass extinction at the Cretaceous–Tertiary boundary, which eliminated the dinosaurs, the ammonites, and many other groups from the fossil record. They considered whether or not certain regions might have been particularly hard hit during this event. Jablonski (1991) suggested that the region a group of species occurred in was a strong predictor of the likelihood that that clade went extinct. Perhaps if mass extinctions are driven by asteroid impact then the regions where these bodies strike are likely to show the highest degree of extinction. However, upon subsequent reanalysis with different types of organisms, Smith and Jeffery (1998) did not find such a pattern. Thus far, studies of biogeographic patterns during mass extinction events have focused primarily on entities within the genealogical rather than the economic hierarchy.

2.8. Conclusions

The hierarchical perspective has profound implications for our understanding of biogeography and paleobiogeography. It suggests that the subdisciplines of historical and ecological biogeography should be redefined to comprise the study of biogeographic patterns in entities of the genealogical and economic hierarchy, respectively. Commonalties between these subdisciplines should revolve around those entities shared between the two hierarchies, especially populations. Further, each entity in each hierarchy can display biogeographic patterns; higher-level entities in the economic hierarchy might show biogeographic patterns depending upon how the debate about coordinated stasis is resolved. Different entities are most prominently influenced by different climatic or geological processes.

Chapter 3

On the Quality of the Fossil Record and What a Paleobiogeographer Can See

3.1. Introduction

Obviously some understanding of the nature of the fossil record is crucial to the discipline of paleobiogeography because it is the filter through which we view all our data, the one true chronicle of the history of life. More than 99.9% of all species that have ever lived are extinct, so if we want to understand something about the history of life, we have to look at the fossil record. The changing biogeographic face of the world will be preserved there. There are, however, also some weaknesses associated with studying biogeographic patterns in the fossil record.

Going back at least to the 19th century, and this is especially apparent in Darwin (1859, 1872), many scientists have commented on how woefully incomplete the fossil record is. And it is clear that although it is our only chronicle of the history of life, it has pieces missing. Darwin actually compared it to a book missing many pages. There are a variety of reasons for this lack of completeness and a subdiscipline within paleontology, called taphonomy, is dedicated to assessing it. One cause is the fact that soft-bodied organisms are absent, except in very rare cases, owing, e.g., to the ravages of scavenging, bacterial decay, and physical degradation. Based on compilations of modern near-shore marine environments, which may be representative of some of the deposits in the fossil record, some 25–30% of the species found there have biomineralized hardparts. The rest of the soft-bodied forms will usually not be encountered in the fossil record.

There are other biases in the fossil record as well. Sediment is not always being deposited everywhere. Certain regions will accumulate more sediment and others less, and then there will be areas that are sites of erosion with no deposition. Obviously, in most cases being entombed is a necessary prerequisite for preservation in the fossil record. The faster an organism is buried, the fewer opportunities there are for the types of processes that break down soft tissue and hardparts, so regions and/or time periods during which there are higher sedimentation rates will tend to be more complete in terms of their preserved fauna. In the case of clastic sedimentary rocks, though not for calcareous deposits, sedimentation in one region is generally produced by erosion somewhere else, so perhaps times of elevated mountain building along with other favorable conditions led to a better fossil record. (In this case it is worth noting that there is a real distinction between marine sections, from which the bulk of the invertebrate fossils come, and terrestrial sections, which preserve the bulk of the vertebrate fossil record.) As a rule, sedimentation is going to require some positive difference between the sediment surface layer and the water level. Above this point there is erosion, and thus no fossils are preserved. Below this level, depending on current velocity and other conditions there might still be erosion, or at least no net deposition of sediments. Potential fossils lying on the sediment-surface/water-level inter-face for an extended period of time without being buried are subject to dissolution, where they are chemically altered by the ocean and broken down.

3.2. Taphonomic Studies

Even after a potential fossil such as a snail shell gets buried there are processes that can affect the types of inferences we might make about it. First, there is the phenomenon of bioturbation. Organisms such as soft-bodied worms, hard-bodied clams, and sea urchins are constantly burrowing into the sediment, churning it up, and eating it, and so possibly mixing up the order in which a set of fossils was laid down. There is also resuspension, when sediments settle and then rise up again, and this can also mix up fossils. Using detailed sampling as well as radiocarbon dating, Flessa *et al.* (1993) and Meldahl *et al.* (1997) studied the temporal magnitude of these types of effects. They found that bioturbation and resuspension can bring 5000-year-old shells to the surface while burying younger ones beneath them. Thus, at least on geologically short timescales, like 5000 years or so, it is difficult to reconstruct the precise temporal sequence of events preserved in a local rock section.

Even after burial there are processes such as diagenesis and compaction that can cause chemical and physical alteration of fossils, and when this happens the structural details necessary for the proper identification of species can be lost. These are only a few among the host of processes that can influence what gets into the fossil record, and we might like some quantitative idea of the degree to which we can say that the fossil record in any one area at any one time, or even in general, might be sufficiently detailed or unbiased to allow us to study certain types of biogeographic patterns. There have been many studies that have looked at this question in a variety of different ways, including one by Valentine (1989), who focused on the Pleistocene and Holocene shallow marine invertebrate fossil record of the California coast. He set out to determine what proportion of the species that are around today that one would expect to be preserved in the fossil record will actually be found there. These would be those marine invertebrates with hard, fossilizable parts such as clams, snails, and sea urchins. He found that more than 80% of these contemporary species were also in the fossil record, proving that it preserves taxa with an amazing degree of fidelity, albeit only for the 25–30% of the species (those with hardparts) that we expect to find there.

3.3. Area and Volume of Sediments

Another pioneering study on the quality of the fossil record was that of Raup (1976), who looked at different time periods in the history of life, such as the Ordovician, the Silurian, and the Devonian. Each of these periods has a different area and volume of rock exposed. There are several reasons for this, including the conditions when they were deposited; the extent of subsequent erosion or further sedimentation, which may have effaced or covered them; the degree to which uplift in a region has brought rocks to the surface; and the duration of these periods. Each of these periods also has a characteristic diversity associated with it. Raup (1976) found a strong statistical correlation between the diversity of a given period and the overall exposed area of rock and/or its volume. However, such a correlation was also found between the area and volume of strata from a given period and the number of paleontologists who had studied these time periods in detail, and he argued that this was a demonstration of the fundamental bias of the fossil record. The preserved diversity is related to what could be preserved, and to the number of people who study it. Perhaps if more paleontologists had worked on certain time periods more species would have been found, or if there were more rocks available to study more diversity might have been recovered.

There may be some circularity in Raup's (1976) conclusions. Perhaps more paleontologists worked on a particular time period because they knew that it had more diversity. Scientists are not likely to knowingly expend effort on something that will not yield interesting results. For instance, each time period has a characteristic number of trilobite species, and researchers who study trilobites generally study them from a single period. There is probably also a correlation between the number of scientists interested in a particular period and the diversity of trilobites in that period, but often with good reason: there are no trilobite researchers studying the Jurassic period because there are no Jurassic trilobites.

Further reasons for believing that Raup's (1976) argument may be circular include the fact that the overall surface area of exposed rock (or its volume) may be related to the amount of time that unit was deposited or the extent of available space that marine invertebrate organisms could occupy on shorelines. Each of these should correlate with the diversity in a realistic, sensible way, which does not depend on biases in the fossil record, but rather is the result of the fact that diversity has varied throughout the history of life because the conditions of life have varied. Still, Raup's (1976) point about potential biases in the fossil record is well taken.

3.4. Phylogenetic Studies

The fidelity of the fossil record has also been studied through the use of cladograms, which make a set of predictions about the patterns of shared relationships among groups of organisms. However, in so doing, they also make a set of predictions about the sequence in which different groups of organisms appeared, based on the topology of the cladogram (Smith, 1994) (see, e.g., Fig. 5). If we assume that the cladogram in Fig. 5 is correct, and that none of the taxa is directly ancestral one to the other, an assumption that will always be true of higher taxa, then monophyletic taxa that share a single common ancestor or have a sister-group relationship must have diverged at the same time (Smith, 1994). That is, the clade comprising B, C and D must have diverged at the same time as the clade comprising group A.

Several paleontologists have taken cladograms generated for different groups of fossil organisms, looked at when these different groups appeared in the fossil record, and compared it with the times predicted by the cladogram. If the cladogram is correct and the fossil record is reasonably accurate, we would expect that the further down a clade appears on the tree, the earlier it would

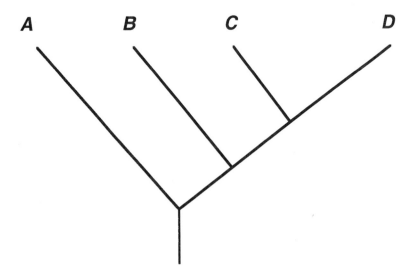

FIGURE 5. A cladogram relating four species.

appear in the fossil record. If the fossil record is reliable, position on the tree, which is usually referred to by the term cladistic rank, should correlate with stratigraphic occurrence; e.g., clades of rank 1 on a particular cladogram should appear first in the fossil record. The stratigraphic first occurrence of a clade can also be ranked, from earliest, lowest rank, to latest, highest rank. Once we uncover the cladistic rank of a taxon and the rank of its stratigraphic first occurrence in the fossil record, we can determine the correlation between these two values using a nonparametric statistic that tests for rank correlation. A good correlation indicates a high-quality fossil record.

The principal studies that utilized this approach [Norell and Novacek (1992), Benton and Storrs (1994), and Benton and Simms (1995)] were all conducted on vertebrate fossils, and traditionally it has been assumed that the vertebrate fossil record is rather poor. However, it turns out that all three of these studies found an excellent correlation between cladistic rank and stratigraphic first appearance, indicating that the overall quality of the record is fairly good.

3.5. Confidence Intervals on Stratigraphic Ranges

Marshall (1990) carried out an important study in which he quantified the extent to which stratigraphic ranges observed in the fossil record represent the true ranges of taxa. This approach justifiably assumes that the fossil record is not a precise record of the history of life. Among other things, this study suggests that the greater the number of horizons a fossil species is known from, the more likely it is that its actual stratigraphic range represents its true stratigraphic range.

3.6. Studies of Sedimentation Rates and Stratigraphic Completeness

In addition to the studies described above, there are many others that have relied on distributions of the individual fossils themselves and/or the rates at which sediments are accumulated in the rock record. Schindel (1980) evaluated various modern environments to determine their characteristic rates of sedimentation and then considered the types of stratal thicknesses that paleontologists needed to resolve to consider certain questions that have been of interest to evolutionary biologists and ecologists. Although his study did not deal with biogeographic questions per se, it is easy to extend his conclusions. It turns out that there have to be very high sedimentation rates and very finely grained sampling to study most short-term biological processes. For instance, processes that occur on the order of 1–10 years are usually preserved in 0.1–1 mm of rock. Processes on the order of 100 years are preserved in 1–10 mm. Other time spans and associated stratal thicknesses are shown in Table 1. Under most circumstances, compaction, bioturbation, and resuspension can

TABLE 1. Stratigraphic Thickness and Temporal Resolution

Time span of interest	Usual thickness of strata for such an interval
1,000 years	10–100 mm
10,000 years	100 mm–1 m

efface anywhere from a few mm up to 10 cm of sediment (Schindel, 1980), which is sufficient to wipe out thousands of years of the fossil record.

A problem that compounds the difficulties of resolution in the fossil record over such short timescales is that even when there are high sedimentation rates, they are often sporadic or intermittent. For example, a river system laying down a large quantity of sediment will generally shift over time, thus eroding what was deposited, which acts against preserving short-term biogeographic patterns and makes the inference of biological processes that operate on short timescales difficult.

Other stratigraphers have considered the completeness of the fossil record, and chief among them are Sadler and Dingus (Sadler, 1981; Dingus and Sadler, 1982; Dingus, 1984), whose studies involved compiling tens of thousands of sedimentation rates. They then plotted these rates against the timescale they were measured for, and used logarithms to reduce the scatter in the data. They found that rates can range over an incredible 11 orders of magnitude, and can also vary with time. As the time span increases, the sedimentation rates fall, owing to the compaction of sediments, the episodic nature of sedimentation, and the fact that over long timescales there tend to be changes in climate and tectonism that alternately facilitate and then restrict sedimentation.

Sadler and Dingus argued that a reasonable way to think of the fossil record is as a set of intervals of sedimentation with juxtaposed gaps. This pattern of preserved record and gaps can be replicated using a simple coin-flipping model of randomly operating sedimentation (Sadler, 1981). They (Dingus and Sadler, 1982) outlined an interesting way of approaching the problem of stratigraphic completeness, arguing that the best way to think of completeness is in terms of the proportion of intervals that are represented by sediment in the section being studied vs. the time span of the whole section. As the timescale of interest gets shorter, there will be a smaller proportion of complete intervals. A section is not complete at the desired limit of resolution if it has any gaps longer than the temporal unit for which completeness is being estimated, and they expressed completeness as a ratio that contrasts the short- and long-term rate of net sedimentation and accumulation (Sadler, 1981; Dingus and Sadler, 1982; Dingus, 1984). Using this perspective, they tried to determine how complete the fossil record is by concentrating on a few key episodes in the history of life, specifically choosing those for which a detailed fossil record might be necessary in order to test certain hypotheses.

One such episode considered by Dingus (1984) was the Cretaceous–Tertiary (K–T) boundary interval that records the extinction of the dinosaurs and the ammonites, and he attempted to determine how complete the boundary sections are. Obviously, a certain degree of completeness is needed to validate sudden scenarios of mass extinction such as those predicted by the supporters of the asteroid-impact theory. An asteroid-impact-induced mass extinction such as the one posited to have occurred at the end of the Cretaceous requires that the extinction happen quite rapidly, on the order of 100 years or so. Is the stratigraphic record well-enough preserved to see events on this time scale? Dingus (1984) considered many K–T sections, including the famous ones at Gubbio, Italy, roughly before, during, and after the K–T boundary. He found that this section is complete only at the level of tens of thousands of years. Thus, it is probably not complete enough to determine whether or not the K–T extinction transpired on a 100-year timescale.

Some corrections to this analysis have been suggested (e.g., Anders *et al.*, 1987), which indicate that the boundary interval may be more complete than Dingus (1984) proposed. Further, this section may not be representative of the fossil record in general, and thus the degree of temporal incompleteness determined by Dingus and Sadler (Sadler, 1984; Dingus and Sadler, 1982) will not always hold true. But, in general, their conclusions seem valid. There may be certain sections where paleontologists can resolve short-time-frame events; however, because of the nature of the fossil record it will be very difficult. The discussions of sedimentation rates and the nature of the fossil record in Schindel (1980), Sadler (1981), Dingus and Sadler (1982), and Dingus (1984) indicate that paleontologists are going to be extremely hard-pressed to come up with sections that allow them to resolve events on a timescale of thousands of years—usually the sharpest resolution attainable is 10,000 years.

3.7. Conclusions

Limitations on resolution make the study of short-term paleobiogeographic patterns and processes in the fossil record difficult. Going back to the discussion of hierarchies in Chapter 2, we see that the lower-level entities in the genealogical and economic hierarchies are going to be inaccessible to paleobiogeographic analysis. Specifically, it will be meaningless to talk about historical paleobiogeographic patterns in any entity below the level of species because biogeographic differentiation within populations (demes) occurs on timescales much shorter than tens of thousands or even thousands of years. In many cases, paleobiogeographic patterns within species will also not be amenable to analysis, and paleobiogeographers will have to focus on paleobiogeographic patterns within clades. Similar constraints hold for ecological paleobiogeography. Paleobiogeographic patterns in regional biotas, if they exist, should be visible in the fossil record. Depending on the nature of communities and how stable they are, they too may be reasonable objects of ecological paleobiogeographic study. However, none of the lower entities in the economic hierarchy is amenable to study with fossil data.

Thus paleobiogeography as a field is restricted to the analysis of only a few entities. However, these entities—clades and regional biotas (and maybe a few others)—persist for long periods of time and occupy broad geographic areas. Therefore, understanding paleobiogeographic patterns within them and the processes that govern these patterns may tell us something important about the history of life—specifically, about how aspects of distribution, evolution, and ecology play out over large areas of the globe and throughout Earth history. As simple extrapolation from patterns and processes at the lowest levels of the hierarchies to the highest is not valid, analysis of paleobiogeographic patterns in the fossil record may be the only way that we can understand the relationship between the evolution of life and the evolution of the Earth, which is why paleobiogeography is important. In a sense, the weakness of the fossil record, i.e., the inability to resolve short-term events, goes hand in hand with its strength, which is that it is the only place where ecologists and evolutionary biologists can hope to see long-term events.

Chapter 4

The History of Biogeography and Paleobiogeography

4.1. Introduction

Biogeography, as many of the disciplines within the biological sciences, especially those once classified as natural history, has a long and rich past. At different times, different ideas or theories held sway, and these governed how natural historians worked as well as what they looked for. Some of these ideas, today's so-called research paradigms, are now seen as dead ends, but, that does not mean that the history of the field has no relevance for modern researchers or that early biogeographic research should be consigned to the dustbin of arcane history. Rather, on the whole, the history of the field illustrates many important things that are of great relevance for modern biogeographers because the major themes of the discipline and the topics that were passionately debated in the 18th and 19th centuries were not very different than they are today. Understanding what these debates were about at an earlier time is important because stripping them of their modernity can teach us things. For instance, part of the debate among current biogeographers is clouded by disagreement over technical and methodological issues. Although these are important, they tend to cause scientists to argue past one another so that they lose sight of the basics that they share. Going back to the core of the discipline can bring all biogeographers to common ground and legitimize various different approaches and topics. Further, it illustrates why certain debates between, e.g., ecological and historical biogeographers are so fundamental that they can never be resolved. But this is not necessarily bad. If we can agree that the rationale behind different approaches is good, then the approach one chooses becomes simply a matter of preference.

My approach to the history of the field will be eclectic, for I am going to emphasize those areas that I believe are of the most relevance to paleobiogeographers. In earlier chapters I discussed the fact that biogeography is best viewed hierarchically. Specifically, the entities in the genealogical hierarchy that historical paleobiogeographers can study are clades and possibly species, and this area of research is well established and productive. Ecological paleobiogeographers can study regional biotas and possibly communities, and this discipline is only in its infancy. It is not clear at this time whether ecological paleobiogeography will develop into a sustainable research program. Therefore, the history of the fields of biogeography and paleobiogeography that is relevant to the analysis of entities in the genealogical hierarchy will be emphasized herein: (1) Is the distribution of animals best explained by dispersal from a single center of origin to a broader region or by invoking the idea that there was once a single, broadly distributed species that later broke up into many different species owing to geologic and environmental changes?; (2) Are the distributions of organisms best explained by their unique ecological factors or by geological events?; (3) What is the role of fossils in shaping scientists' views on biogeography?

4.2. Preevolutionary Biogeographic Views

4.2.1. Centers of Creation, Vicariance, and Dispersal

The initial approaches to explaining biogeographic patterns, first developed in the 16th and 17th centuries, were based on biblical doctrine. Then, the belief was that after the great flood the animals disembarked from Noah's Ark on Mount Ararat and dispersed around the globe (Kinch, 1980; Mayr, 1982; Browne, 1983). Natural historians at that time would have predicted that the species of the New World would be the same as those known from Europe. However, by the 18th century, the age of exploration had brought many new species back to Europe. From a scientific standpoint, some of the most productive of voyages of that period, were those led by Captain Cook on the *Endeavor*. During a three-year voyage with Cooke, from 1768 to 1771, Sir Joseph Banks collected over 1000 new species of plants (Watkins, 1996; Brown and Lomolino, 1998) (for further discussion of Banks see Chapter 11).

So many new species were discovered during the 18th century that it no longer seemed plausible that they could all have fit aboard an ark (Mayr, 1982;

Browne, 1983), and dispersal from the ark to explain the geographic distribution of plants and animals was no longer held to be valid. In addition, natural historians came to realize that there were several major regions of the Earth, each with its own largely unique flora and fauna (Kinch, 1980). This was the fundamental biogeographic pattern that needed to be explained:

> That different regions of the globe are inhabited by entirely distinct animals and plants is a fact which has been familiar to all naturalists since Buffon first pointed out the want of specific identity between the land quadrupeds of America and those of the Old World.... the extent of this parceling out of the globe amongst different nations, as they have been termed, of plants and animals, the universality of a phenomenon so extraordinary and unexpected, may be considered as one of the most interesting facts clearly established by the advance of modern science. (Lyell, 1832, p. 66)

Linnaeus, who was the father of systematics, was one of the first natural historians to articulate a biogeographic theory to explain this pattern that did not rely on dispersal from Noah's Ark. However, he did believe that his study of biogeographic patterns was part of a great endeavor to interpret God's wisdom. He posited that each species was created as a single pair of organisms on a small, mountainous island surrounded by a great flood (Kinch, 1980; Browne, 1983). Each species was created in a particular climatic belt that corresponded to a zone of elevation on the island, and to which it was perfectly suited; as the flood waters receded, different species dispersed outward from this island to colonize their preferred environments (Browne, 1983).

The great 18th-century French naturalist Buffon challenged Linnaeus' explanation (Mayr, 1982), recognizing that if every species was perfectly suited to an environment, it was extremely unlikely that it could disperse over great distances, for it would need to cross many inhospitable terrains during the migrations. Instead, he asserted that species originated in the regions that they currently inhabit (Mayr, 1982; Browne, 1983); i.e., that there were multiple creations, with different floras and faunas created in separate regions (Kinch, 1980). Further, Buffon (1756) believed that the natural world makes the species. "The Earth makes the plants; the Earth and the plants make the animals" [Buffon (1749–1804) in Mayr (1982, p. 441)]. This pithy statement is a clear argument for the validity of biogeographic studies. Nelson (1978) codified Buffon's ideas on biogeography into Buffon's law (already formulated as such in the 19th century), which stated that different areas have different species. Brown and Lomolino (1998) recodified it in terms of the prediction that environmentally similar but isolated regions have distinct assemblages of mammals and birds.

Willdenow and Zimmermann were two other important early biogeographers who championed a view similar to Buffon's. Zimmermann believed that if countries have similar species then at one time they must have been connected. In referring to Ceylon and Great Britain he invoked former connections with Asia and Europe, respectively (Mayr, 1982).

Thus, shortly after the establishment of the discipline of biogeography, explanations of biogeographic patterns had begun to fall into one of two camps. Although it might be unfair to polarize these two viewpoints, the original fundamental disagreement in biogeography focused on whether species originated in a single region and spread out or originated in a single region and remained there. The former approach relies on dispersal, and the latter denies it. The question of whether dispersal of taxa or their *in situ* origin best explains biogeographic patterns is not by any means settled. In fact, it constitutes one of the fundamental debates in modern biogeography [see, e.g., Rosen (1978, 1979), Nelson and Platnick (1981), Brooks (1985, 1988), Wiley (1988*a*,b), Brooks and McLennan (1991), and Brown and Lomolino (1998)]. However, I am going to argue later in this chapter and in Chapter 6 that there is a means of resolving this debate.

By the late 18th and early 19th centuries the viewpoint that Buffon had advocated largely held sway. Geographic regions were held to be unique entities made up of a characteristic flora and fauna. These species were believed to be bound to one another and to their geographic locale (Browne, 1983). The connection at this time between the Earth and its biota was believed to be so strong that Willdenow, the mentor of the famous biogeographer von Humboldt (who will be discussed shortly) said

> ...by the history of plants we mean a comprehensive review of the influences of climate upon vegetation, of the changes which plants most probably have suffered during the various revolutions this earth has undergone, of their dissemination over the globe, of their migrations, and lastly of the manner in which nature has provided for their preservation. (Willdenow, 1805 in Browne, 1983, p. 39)

The belief that this connection was strong meant that the study of the geographic distribution of animals and plants might reveal fundamental natural laws (Browne, 1983). In von Humboldt's early 19th-century work we see evidence of one of the first biogeographers who approached the study of the geographic distribution of organisms in this way—as the pursuit of natural laws. For instance, he noted that different climates had typical associations of plants, and he argued that this could be clearly seen while ascending a mountain. He did not necessarily hold that the tropical regions of different areas shared the same genera, but rather that the overall form of tropical plants would be the same everywhere (Browne, 1983). Actually, according to Nelson (1978), Humboldt's (1816) ideas on distribution can be summarized with the precept that there are no animals in common between the equatorial regions of the New and Old Worlds (the Americas and Europe and Africa). The distribution of animals, in Humboldt's (1820) view, would be due to climatic and geological causes that are unknown (Nelson, 1978).

4.2.2. Augustin de Candolle

Although the natural scientists discussed above as well as many other early biogeographers made substantial contributions to the theoretical development of the field, none of their work equals that of Augustin de

Candolle. In three works de Candolle (1817, 1820, 1821) developed the notion that there is a distinction between the factors that determined the distribution of species on a small scale and those that influence global-scale patterns (Browne, 1983). Factors operating on the small scale include temperature, light, and the ecology of organisms, and these control what he called the stations of organisms. Factors operating on the large scale were related to the fact that the organisms of the world (specifically the plants of the world) were divided up into provinces, which he called habitations. The distribution of these habitations was also governed by laws (Browne, 1983). Thus, for de Candolle, local climate and what we now call ecology could not, alone, explain the distribution of organisms; there are also historical, biogeographic patterns of differentiation. He argued that

> It might not, perhaps, be difficult to find 2 points in the United States and in Europe, or in equinoctial America and Africa, which present all the same circumstances: as, for example, the same temperature, the same height above the sea, a similar soil, an equal dose of humidity, yet nearly all, perhaps all, the plants in these two similar localities shall be distinct...circumstances, therefore, different from those which now determine the stations have had an influence on the habitations of plants. (de Candolle, 1820 in Lyell, 1832, p. 68)

The recognition that both local climate and ecological factors as well as historical biogeographic factors influence biogeographic patterns is extremely important. Clearly animals and plants are adapted to climate, and as climate changes, they tend to track their preferred environments. However, these habitat preferences alone cannot account for large-scale biogeographic patterns. Thus, de Candolle's view is in some ways similar to the one taken in the section of this book that emphasizes hierarchies (though he was only concerned with what we would now call the entities of the genealogical hierarchy). His view implies a belief that different processes operate and different patterns prevail at different levels of the genealogical hierarchy. One cannot simply extrapolate lower-level processes influencing biogeographic patterns in populations to explain higher-level biogeographic patterns in, e.g., species and clades because there is a difference between the processes that control the distributions of individual species and those that influence the way various groups of species are distributed:

> This fact leads us to the idea...that stations are determined uniquely by physical causes actually in operation, and that habitations are probably determined in part by geological causes that no longer exist today. According to this hypothesis one may easily conceive why plant species that are never found native in a certain area will nevertheless live there if they are introduced. (de Candolle, 1820, in Nelson, 1978, p. 284)

De Candolle's belief that there is a disjunction between small- and large-scale patterns was an extremely prescient one. Like the perspectives of Linnaeus and Buffon, it anticipated an important debate in biogeography in particular and evolutionary biology in general: Specifically, is it possible to extrapolate small-scale phenomena to larger scales? As I noted in Chapter 2, some evolutionary biologists have argued that there is a distinction between microevolution and macroevolution such that processes at one scale within lower-level entities of the genealogical hierarchy cannot be freely extrapolated to explain patterns within large-scale entities of the same hierarchy (e.g., Eldredge, 1979; Gould, 1980; Vrba, 1980). (Others have argued against this notion [e.g., Dobzhansky (1937), with hesitancy, and more recently Charlesworth *et al.* (1982)]. Similarly, the same question of whether or not small-scale phenomena can be extrapolated to larger scales has been debated in biogeography. Some researchers, e.g., MacArthur and Wilson, 1967 supported the extrapolationist view. Their ideas on biogeography were based on principles of population ecology and genetics, and they did not see any real difference between population genetics and biogeography (Patterson, 1983). However, because processes influencing the distributions of populations cannot be extrapolated to explain the distributions of groups of related species their approach lacks validity except at the population level.

4.2.3. The First Paleobiogeographers

Although the aforementioned biogeographers such as Linnaeus, Buffon, von Humboldt, and de Candolle were very influential in the field, their impact on the particular discipline of paleobiogeography is associated primarily with their work in biogeography in general. They were not interested in fossil forms and were concerned only with present-day biogeographic relationships. Among the first paleobiogeographers were de Candolle's son Alphonse and Adolphe Brogniart, both of whom were paleobotanists whose most important works were published in the 1840s. They both believed that life appeared as a single primitive population distributed over the globe, and that gradually this population fragmented into many groups of species such that there was a vector throughout Earth history from one to many biological provinces (Browne, 1983). The driving force behind this fragmentation would have been a change in the Earth's environment from uniform to diverse. Early on, they marshaled evidence in support of this contention from the Eocene plant fossil record of Europe; however, when new fossil material came in from Australia and South America, they and other natural historians began to believe that there were as many paleobiogeographical provinces as there were modern biogeographical provinces (Browne, 1983).

By the 1840s the belief that there was a single uniform population began to fade, or it was argued that it was a phenomenon of the Silurian period (Browne, 1983). This is actually quite interesting, for what was referred to as the Silurian in the early 19th century is partly equivalent to the modern Cambrian. The major animal groups first appeared and then radiated during the late pre-Cambrian and Early Cambrian, between roughly 570 and 530 million years ago. In the late pre-Cambrian, the Earth's plates were amalgamated into a single supercontinent, referred to as Rodinia, which was largely submerged beneath a shallow seaway (Hoffman, 1991; McKerrow *et al.*, 1992; Dalziel, 1997). Perhaps in the late pre-Cambrian there really were only a few animal populations, uniformly distributed across this vast region. Then, this supercontinent fragmented, separating faunas and facilitating their diversification (Dalziel, 1997; Lieberman, 1997).

Up to the early to mid 19th century two different perspectives on biogeography had been put forward. One, first-posited by Linnaeus, argued that species spread out or dispersed from a single center of origin. The other, evident in the works of Candolle fils and Brogniart, argued one or a few species originally had a very broad range, which was later fragmented, producing many more species occupying smaller geographic regions. These two different perspectives cannot actually be mutually exclusive. In the case of species with broad ranges, a question that might be posed is how did that range originally come to be broad. Unless a miraculous explanation is invoked, it is hard to imagine that the range became broad without some expansion from an originally more restricted region. Similarly, dispersal out of a single narrow center of origin begs the question of how that original distribution came to be narrow in the first place. Could it be because a primitively widespread species fragmented into several more narrowly distributed species?

By invoking this possible logical inconsistency in the works of early biogeographers, I do not mean to single out these great natural historians for censure or blame. Rather I want to point out that the legacy of this debate is still with us. Indeed it is the raison d'etre of the discord between the so-called vicariance biogeographers, whose theoretical stance better fits that of Buffon, Zimmermann, Candolle fils, and Brogniart, and the dispersalist biogeographers, who have more in common with the views of Linnaeus.

4.2.4. Charles Lyell and Resolution of the Vicariance/Dispersal Debate

Charles Lyell, in addition to being one of the most important geologists to have ever lived, also made several contributions to the field of paleobiogeography. He recognized that understanding why certain groups of animals and plants were present in certain regions, and absent from others, was extremely important for science. "Next to determining the question whether species have a real existence, the consideration of the laws which regulate their geographical distribution is a subject of primary importance to the geologist" (Lyell, 1832, p. 66). The biogeographic phenomenon that most intrigued Lyell and his contemporaries was that the world was broken up into a several biogeographic regions. Lyell (1832) was a profound admirer of the views of de Candolle père on the existence of stations and habitations of organisms, and saw this as offering a potential explanation for why the world was divided up into different biogeographic regions. His own explanation was a very important contribution to the debate about whether vicariance or dispersal most strongly influences biogeographic patterns in organisms, though it was largely ignored.

Lyell (1832) believed that Earth history unfolded in a series of cycles rather than in a single, linear trajectory (Browne, 1983; Gould, 1987), a philosophical approach that was partly derived from his study of the fossil record. He believed that during each epoch, species expanded and contracted their boundaries as geographical barriers were removed or formed, i.e., patterns of both dispersal and subsequent range contraction occur, albeit at different times. He specifically emphasized patterns of dispersal, claiming that if no topographic or ecological barriers to dispersal existed, species could spread around the globe. He in fact stated that there was a general law governing the geographical distribution of organic beings, "the limitation of groups of distinct species to regions separated from the rest of the globe by certain natural barriers" (Lyell, 1832, p. 87). The geologist Forbes (1846) reiterated this viewpoint.

Lyell (1832) also combined these views on dispersal with an emphasis on vicariance:

> However equally in this case our original stocks were distributed over the whole surface of land and water, there would nevertheless arise distinct botanical and zoological provinces, for there are a great many natural barriers which oppose common obstacles to the advance of a variety of species. (Lyell, 1832, p. 125)

Oceans and climatic belts would represent important barriers. He described how, in this manner, an originally broadly distributed ancestral species might differentiate into a more narrowly distributed descendant species.

Lyell's contributions, and those of Forbes, reflected an emphasis on deep time, and thus had a critical dependence on an appreciation of the fossil record. Lyell and other paleontologists recognized that the geographic ranges of animals and plants were dynamic entities that grew or shrank with geologically mediated changes. This served to dislocate, reconnect, mix, and isolate organisms (Browne, 1983). He also concluded that distributions at earlier times were no more simple than those of the present day, a viewpoint that Agassiz (1842) reiterated. Lyell's views on dispersal were probably adapted from Prichard's (1826); however, the latter believed that it was supernatural creation that controlled the division of the Earth into separate biogeographic regions. By contrast, Lyell (1832) believed that these patterns could be produced by both the dispersal of species, which was limited by barriers (Kinch, 1980), and by what we would now term vicariance.

Lyell further emphasized that there was a continuous interaction between the biological and the geological realm such that the two cannot be separated, and for him the best way to study biogeography was to see how the animate world is affected by the inanimate. He thought that this could be ascertained by looking at the positions of groups of species and how these have changed owing to migrations and changes in physical geography. He also argued that scientists could not just look at the present-day configuration of the continents and understand the geographic distribution of organisms. Instead, he insisted, they needed to understand the history of those continents. This view was originally forcefully stated by Lyell, but it was perhaps expressed most articulately by Whewell (1840), "the past has been a series of events connected by historical causation, and the present is the last term of this series" (quoted from Browne, 1983, p. 105).

As discussed above, Lyell (1832) believed in the importance of barriers in structuring biogeographic patterns. However, he also recognized that barriers can be removed as a result of geological change. This illustrates one of the differences between Lyell and his predecessors in the field of biogeography. The real distinction between Lyell and earlier biogeographers such as von Humboldt, de Candolle père, and Buffon was that Lyell took a dynamic, historical approach, emphasizing geological processes that influenced biogeographic patterns while the others viewed biogeographic patterns as static events divorced from the geological processes that might have engendered them (Browne, 1983). Lyell was the world's first paleobiogeographer, and his approach, which was a result of his vast experience with patterns in the fossil record, makes good sense: if we are to reconstruct the history of biogeographic patterns within organisms it is necessary to turn to the fossil record because it contains the one true chronicle of the history of life.

It was Lyell's unique perspective as a geologist that enabled him to come up with the theoretical framework that can resolve the current debate between vicariance and dispersalist biogeographers over which phenomenon, vicariance or dispersal, most influences biogeographic patterns. He recognized that geological processes facilitate both the contraction of the ranges of species relative to their primordial forms, now referred to as vicariance, and also range expansion of species relative to their primordial forms, now referred to as geodispersal (see Chapter 6). In other words, geological processes can produce congruent vicariance and range expansion, contra modern biogeographers who emphasize only vicariance. Other paleontologists have championed this view (e.g., Forbes, 1846; Simpson, 1965; Hallam, 1967; McKenna, 1983). However, the idea that geological processes can facilitate both range expansion and range contraction has not been integrated into historical biogeography, which is primarily dominated by scientists who study only extant organisms. These biogeographers must, of necessity, study a single time slice of a group rather than its entire history, and this view is incomplete because "the key to the distribution of any group lies in the geographic configuration of that epoch in which it made its first appearance" (Gadow, 1913, p. 13) as well as in all subsequent geographic configurations.

It is Lyell's unique geological and paleontological perspective that make his writings critical for biogeographic theory. Specifically, he demonstrated the relevance of paleontology and a deep-time perspective to biogeography, and in so doing he founded the discipline of paleobiogeography. He also emphasized the enormous influence of geological processes on still-living organisms. This viewpoint was taken up by Darwin (1839) in his consideration of the formation of coral atolls and many other topics.

Biogeographers studying the modern biota appreciate the role that geological processes play in motivating biogeographic patterns (e.g., Croizat et al., 1974; Nelson, 1976; Brooks et al., 1981; Wiley, 1981, 1988a,b; Wiley and Mayden, 1985; Cracraft, 1988; Kluge, 1988; Mayden, 1988; Brooks and

McLennan, 1991). However, it is still crucial, if biogeography is to become more complete, for paleontologists to become more integrated into the field. In other words, paleobiogeography and biogeography must come to imply two aspects of the same discipline rather than two different disciplines. However, in order to accomplish this, paleobiogeographers must come up with a set of rigorous analytical methods to match those of biogeographers. As there are processes of vicariance and dispersal that paleobiogeographers can track over time, they require analytical methods that take these two phenomena into account in order to analyze their data. Methods that have been adapted for this are discussed in greater detail in Chapter 9.

Although the works of the early biogeographers that I have been discussing were compatible with the fact that life had evolved, their authors did not actually believe in evolution. Rather, they thought that each region with its own diverse fauna represented a separate center of creation. Lyell (1832) concurred in this notion, and further believed that species were added to the Earth by the creator at different times. However, he also believed that there was some relationship between the geographic distribution of species and their distribution in the fossil record. This is concisely stated in his famous aphorism, "As in space, so in time" (Lyell, 1832, quoted from Browne (1983, p. 149). Forbes (1846) also saw a relationship between the geographic distribution of species and the time when these species appeared. However antievolutionary the stance of Lyell and other early biogeographers, eventually integrating the relationship between the temporal appearance of species and their geographic distribution led to the demonstration that life had evolved. How this came to be will be discussed at length below.

4.2.5. Charles Darwin and the *Voyage of the Beagle*

Charles Darwin, a profound admirer of Lyell, also recognized the association between the geographic distribution of species and the manner in which different species appeared through time. This is first evidenced in his *Voyage of the Beagle* (Darwin, 1839) sometimes known as the *Journal of Researches*, which was written as a personal narrative describing a set of biological and geological observations based on a circumnavigation of the globe. He later claimed that this trip determined his whole career.

Darwin (1839) described several biogeographic phenomena that are of great relevance to the history of the field of biogeography, one of which involved the geographic distributions of two species of large flightless birds or "ostriches" from South America. He found that one species of ostrich, the South American rhea, was replaced by another similar species, which he called the Petite Avestruz (Darwin, 1839, p. 108), in southern South America, such that their geographic ranges did not overlap, and that the boundary was just south of the Rio Negro. This discussion was not, at the time of publication, tied by Darwin or anyone else to the concept that life may have evolved. However, it became important when he later combined this observation with one that was initially unrelated, based on his studies of the fossils of the South American continent.

When Darwin looked at some of the Tertiary mammal fossils of the Pampas, what are now called the macrochaeniids and the glyptodonts, he decided that these animals were different from, yet closely related to, the present-day guanacos or llamas and the armadillos of the same region (Darwin, 1839, p. 162). Although this conclusion seems unrelated to his observation concerning the ostriches, he recognized their connection in a novel way some time after *Voyage of the Beagle* was published, and this is documented in his originally unpublished notebooks. Thankfully we can see this insight because there is a rich tradition of research by historians of science on these notebooks, which provide us with a glimpse as to how Darwin's ideas about evolution evolved over a 20-year period—after his return from the voyage of the *Beagle* but prior to the publication of *On the Origin of Species*.... The relevance of his notebooks to the field of biogeography will be discussed in greater detail below.

Darwin (1839) also recorded many other observations on the geographic distributions of animals that are of interest to biogeographers. For example, he recognized that

> ...whole series of animals, which have been created with peculiar kinds of organization, are confined to certain areas; and we can hardly suppose these structures are only adaptations to peculiarities of climate or country; for otherwise, animals belonging to a distinct type, and introduced by man, would not succeed so admirably, even to the extermination of the aborigines. (Darwin, 1839, p. 165–166)

Like de Candolle père, Darwin recognized that climate alone could not explain the distribution of species on the globe, and that there was some other reason why certain species were found in particular areas. He might not have realized it at the time, but part of the reason is the existence of topographic barriers. He did discuss the significance of such barriers as boundaries of distribution for some groups of animals. For instance, he cited (Darwin, 1839, p. 249) examples from Chile where different animals are found on different sides of the Cordillera, and noted that species on the eastern side are similar to species in Patagonia, which lies even further to the east.

Many of the examples that Darwin (1839) cited that have proved to be of considerable interest to later evolutionary biologists, biogeographers, and historians of science involve the fauna of the Galapagos Islands. The most important of these from the standpoint of biogeography concern the distributions of different species on the various islands of the archipelago. This fact originally surprised Darwin, "...it never occurred to me that the productions (meaning species) of islands only a few miles apart, and placed under the same physical conditions, would be dissimilar" (Darwin, 1839, p. 287). He found it to hold true for mockingbirds, and, based on discussions with other people, he was made aware of the fact that the large tortoises of the Galapagos were slightly different from island to island (Darwin, 1839, pp. 279,

287). Eventually, this fact of different species on different islands came to strike him with wonder (Browne, 1983).

The most frequently cited example of this phenomenon involves the so-called Darwinian finches of the Galapagos. At the time of his travels, Darwin thought that there was only a single species of finch on the islands, some of which had thicker beaks than others (Darwin, 1839, p. 287), so he collected specimens of finches without taking note of the island on which they were found. Only after his return to England did the great ornithologist John Gould, who Darwin had given his bird specimens to, point out his error. Gould concluded that there were actually several different species of finches, each of which was later found to be unique to a particular island in the chain (Grinnell, 1974). This was to have an important influence on Darwin's views, and it is considered more fully below in the section on the Darwinian notebooks.

On a broader spatial scale, Darwin also recognized that the birds of the Galapagos, albeit unique species, were similar to birds of South America in their overall type. He speculated that perhaps this was because the creative power (relating to the creation of species) acted according to the same law over a wide area (Darwin, 1839, p. 287). On the face of it, it would be hard to explain, if species were created independently, how those on an island chain hundreds of kilometers off the coast of South America would resemble others from the South American mainland, yet at the same time differ from them. It is clear now how all of these facts could be easily explained within an evolutionary framework, though Darwin did not originally see them that way. His basic belief was that the species of the Galapagos had been created in the central islands and later dispersed out a certain distance (Darwin, 1839, p. 282).

4.3. Evolutionary Biogeography

Among the greatest scientific developments in the field of biogeography was the demonstration that the facts of geographical distribution supported the notion that life had evolved. In the middle of the 19th century many natural historians believed as did Louis Agassiz (1842), that species were supernaturally created at the place and for the place in which they live. For the science to progress, this reliance on supernatural explanations had to be abrogated.

An important early biogeographer who seemed to recognize an association between what we now call evolution and biogeographic patterns was von Buch, who collected the flora of the Canary Islands and established its relationship with African floras. Although he operated within what is traditionally believed to be a preevolutionary framework, it is clear that a nascent idea in his work was that geographic differentiation leads to evolutionary change. He drew upon some of the ideas that de Candolle père developed about habitations and stations as well as Lyell's ideas about dispersal. He recognized that dispersal ability is related to the ecological characteristics of species (their stations). Further, evolutionary divergence, which leads to the separation of floras and faunas (habitations) is related to barriers that isolate populations geographically and prevent interbreeding:

> On the continents the individuals of a genus are dispersed far and wide and, due to the differences of station, food, and soil, form varieties which, at this distance (that is, geographically isolated), not intercrossing with other varieties and therefore not reverting to the parent type, eventually became constant and separate species. Then if, by chance, in other directions, they happen to meet with another variety which has been similarly modified, the two, being very different species, can no longer intermix. (von Buch, 1825, in Kottler, 1978, p. 285).

Von Buch (1825) essentially laid out a theory of evolution by common descent mediated by geographic barriers that isolate populations of species (Sulloway, 1979). This theory relied not so much on geologically mediated changes leading to geographic barriers, but rather on chance dispersal by organisms over some preexisting barriers. He noted that once some seeds passed over a barrier that previously represented the limits of that species range, over time the result will be "a distinct species which diverges from its parent form in proportion to the length of time it has been isolated" (von Buch, 1825, in Kottler, 1978, p. 286). He commented on a genus of plants with species in the Canary Islands, and remarked that "one finds himself naturally led to consider all the species of this genus as descendants from a common ancestor. These different species are almost never found in the same place but each is limited to its own valley or a particular district" (von Buch, 1825, in Kottler, 1978, p. 286).

Von Buch's great insights on how biogeographic patterns are driven by evolutionary divergence, mediated by geographic isolation, appear to be limited to this one bibliographic source, but they were clearly prophetic. He relied on isolation to produce evolutionary change. Such evolutionary change will be reflected as biogeographic differentiation in several groups. Further, isolation arises from dispersal over geographic barriers. Although von Buch's views have a strong Lamarckian component, to the extent that he held that the external environment shaped species, he clearly deserves credit for being an early if not the earliest evolutionary biogeographer. Both Wallace and Darwin read his work, and it is clear that they were influenced by his ideas, as isolation and geographic barriers figure prominently in their theories of evolution, as outlined more fully below. Further, first Darwin, and later Wallace, came to rely on the notion of dispersal facilitating isolation.

Chambers (1844) was another important early natural historian, who related the distribution of organisms to a broader theory of evolution. He argued publicly that separate biological faunas had evolved *in situ*. They were not separate biological creations.

4.3.1. Alfred Russell Wallace

The first scientist whose published work unambiguously recognized the connection between patterns of the geographic distributions of animals without relying on Lamarckian mechanisms of evolutionary change was Alfred Russell Wallace. He was the first to assimilate Lyell's aphorism on the concordance between the spatial and temporal distribution of fossils and integrate it into an evolutionary theory. Thus, he is the first true evolutionary biogeographer. This is not to say that Wallace's views were completely concordant with those of Lyell. In fact, in many respects, they differed greatly, as will be discussed more fully below.

Wallace's background as a scientist and natural historian included extensive experience collecting rich and diverse faunas in the tropics of South America and Asia. The discovery and analysis of these relatively unknown and rich faunas clearly had an important influence on the development of his biogeographic views relating to the origin of species and evolution. Further, he was greatly influenced by the career and writings of von Humboldt. What is perhaps most amazing is that many of Wallace's most important insights into biogeography and evolution came during arduous collecting trips in what most people would have considered extreme environments. Obviously, he found inspiration in these voyages, and in so doing he enriched science.

Wallace's earliest major collecting trip began in 1848 when he joined with the great entomologist Henry Bates to collect the fauna of the Amazon, the same region von Humboldt had visited some 50 years earlier. One of their objectives was to look at the relationship between patterns of affinity and patterns of geographic distribution, in closely related species in order to see how species arise (Brooks, 1984). In the course of their travels they discovered that animals tend to live in small, local groups that occupy territories with distinct boundaries often formed by rivers (Brooks, 1984). Specifically,

> ...the Amazon, the Rio Negro and the Madeira (rivers) formed the limits beyond which certain species never passed. The native hunters are perfectly acquainted with this fact, and always cross over the river when they want to procure particular animals. (Wallace, 1852, from Brooks, 1984, p. 35)

Although this South American expedition was very important for Wallace's intellectual development, perhaps his most productive period, in terms of innovative ideas, was the 8 years he spent collecting samples in the Malay Archipelago from 1854 to 1862. His greatest theoretical contribution, though sadly it was completely ignored by his contemporaries (Brooks, 1984), until it was cited in Darwin (1872), was probably the paper entitled "On the Law which has regulated the introduction of new species", Wallace (1855) in which he stated that the Earth has gone through a series of gradual changes, and so too has organic life. He went on to codify the distribution of organisms in space at the time he was writing, their geography, as well as their distribution through time, their geology (Brooks, 1984). Thus, his contributions to evolutionary and biogeographic theory incorporated an appreciation of the fossil record. Although he did not consider biogeographic patterns in the fossil record, and thus cannot be considered a paleobiogeographer, he recognized the importance of understanding the history of life for formulating a theory of evolution.

It is worth discussing in detail how Wallace met the expectations of Lyell's aphorism about space and time, and molded them into a superb theory of evolution. First, he claimed that there was a law that "every species has come into existence coincident both in time and space with a preexisting closely allied species" (Wallace, 1855, quoted from Brooks, 1984, p. 172). Further, he felt that this

> ... connects together and renders intelligible a vast number of independent and hitherto unexplained facts. The natural system of arrangement of organic beings, their geographical distribution, their geological sequence .... Granted the law, and many of the most important facts in Nature could not have been otherwise, but are almost as necessary deductions from it, as are the elliptic orbits of the planets from the law of gravitation. (Wallace, 1855, in Brooks, 1984, p. 79–80)

Beyond just making this poetic statement, Wallace (1855) put together facts from geology and biogeography to make a case for evolution. In terms of geography, he held that there was a hierarchy of distribution from large groups such as classes and orders, which were generally spread over the whole Earth, to smaller ones such as families and genera that were confined to a part of the globe (see Chapter 2). He also recognized that when a diverse group is found, it is likely that closely related or allied species will be found in the same locality. That is, "the natural sequence of the species by affinity is also geographical" (Brooks, 1984, p. 73). More succinctly, geographical distribution and evolutionary relationship are correlated, and he recognized analogous geological patterns. He noted that "the phenomena of geological distribution are exactly analogous to those of geography. Closely related species are found associated in the same beds, and the change from species to species appears to have been as gradual in time as in space" (Wallace, 1855, in Brooks, 1984, p. 75). For instance, he concluded that species of a genus occurring in the same geological time period are more closely related than those separated in time. Further, he realized that related species do not occur in far-flung geographical regions without also occurring in intermediate locales, just as species do not appear in the fossil record at one time period and then at a much later period.

Among the biogeographic data adduced to support his view of evolution was the fact that diverse faunas and floras tended to be geographically isolated:

> If we now consider the geographical distribution of animals and plants upon the earth, we shall find all the facts beautifully in accordance with, and readily explained by, the present hypothesis (evolution). A country having species, genera, and whole families peculiar to it will be the necessary result of its having been isolated for a long period, sufficient for many series of species to have been created on the type of pre-existing ones. (Wallace, 1855, in Brooks, 1984, p. 75)

Wallace (1855) also demonstrated a familiarity with Darwin (1839), and commented on some of the observations in the *Voyage of the Beagle*, specifically those pertaining to the Galapagos Islands. Here, he reiterated his deep insight into the role that geographic isolation played in motivating evolutionary change. However, he took this insight one step further by using the Galapagos Islands to outline how a single broadly distributed ancestral species can become broken up into several more narrowly distributed descendant species:

> Separate islands have different species now. Originally they had the same species from which differently modified prototypes were created, or the islands were successively peopled from each other, but new species have been created in each on the plan of the pre-existing ones. (Wallace, 1855, in Brooks, 1984, p. 101)

This is an early description of the process of vicariance, which forms the major component of modern cladistic attempts to reconstruct biogeographic patterns. de Candolle fils and Brogniart had outlined such a process of vicariance earlier, but Wallace (1855) was the first to integrate it into an evolutionary framework.

What is generally conceived to be Wallace's most important contribution to biogeography is his elucidation of the great disjunction between the flora and fauna of the islands of Bali and Lombock (e.g., Van Oosterzee, 1997). Only narrow straits separated these islands, which later came to represent the eponymous Wallace Line that demarcated the eastern- and southernmost boundary of the Asian zoological province from the northern- and western-most boundary of the Australian zoological province. In retrospect, it seems unfortunate that this should be considered Wallace's most important biogeographic discovery. He thought he had delineated the boundaries between two great biogeographic provinces—the Australian and the Asian—and he certainly deserves immense credit for this, as it entailed a significant amount of grueling collecting. He also demonstrated how the zone of interdigitation between provinces could be narrow, but natural historians had recognized that the world is broken up into separate centers of creation or zoological provinces since the 18th century. Further, the one great source where he elucidated the facts of this faunal discontinuity in the Malaysian Archipelago (Wallace, 1869) contains no explicit mention of the relevance of this phenomenon to a broader theory of evolution. Thus, his was a great factual discovery, not a theoretical one.

This factual discovery was inspired by the work of Sclater (1857), who, in an essay on birds, divided the Earth into six great ornithological regions, each characterized by a distinct avian fauna. Each of these regions, according to Sclater represented a separate center of creation. He also deemphasized the role of migrations as an explanation of these centers, although it figured prominently in that context in the works of Prichard (1826), Lyell (1832), and Forbes (1846), because he believed that each species was designed to be perfectly adapted to the environment it was found in (Kinch, 1980). The subdivision of the Earth's biota into a series of distinct regions had a long tradition prior to Sclater's work, as I discussed above; however, his codification and elucidation of these regions was particularly rigorous.

Wallace (1863) believed that Sclater's regions applied to many other groups of organisms as well.

In reality, Wallace stumbled upon a pattern that he was not looking for. He hoped to discover fine-scale patterns of distribution, to look at how new species arise but, instead, in the discontinuity between Bali and Lombock, Wallace's Line, he found coarse-scale patterns (Brooks, 1984). The facts he uncovered were not what he expected and did not relate, at least directly, to the issue of how species originated. Still, he used them to refute earlier viewpoints, such as those of Lyell (1832), which argued for the successive creations of different species. Wallace posed the question "why are not the same species found in the same climates all over the world?" (Wallace, 1857, in Brooks, 1984, p. 160), which is essentially a restatement of the distinction de Candolle père made between the stations or climatic tolerances of species and their habitations, or where they occurred. He went on to attack Lyell (1832):

> ... the general explanation given (that of Sir Charles Lyell) is, that as the ancient species became extinct, new ones were created in each country or district, adapted to the physical conditions of the district.... [However, if this were true, we should find] a general similarity in the productions of countries which resemble each other in climate and general aspect while there shall be a complete dissimilarity between those which are totally opposed in these respects. However, the islands of Borneo and New Guinea are very similar in area and climate, yet totally differ in productions. By contrast, Australia and New Guinea are very different in their physical conditions. One is near the equator, the other is near the tropics, yet the faunas of the two, though mostly distinct in species, are strikingly similar in character.... We can hardly imagine that the great variety of monkeys, of squirrels, of Insectivora, and of Felidae, were created in Borneo because the country was adapted to them, and not one single species given to another country exactly similar, and at no great distance. We can hardly help concluding, therefore, that some other law has regulated the distribution of existing species than the physical conditions of the countries in which they are found, or we should not see countries the most opposite in character with similar productions, while others almost exactly alike as respects climate and general aspect, yet differ totally in their forms of organic life. (Wallace, 1857, in Brooks, 1984, pp. 160–161)

Thus, Wallace skillfully used biogeographic facts that had been elucidated by natural historians who believed in special creation, such as the distinction between the habitations and stations of organisms, to demonstrate that special creation, at least Lyell's (1832) version of it, could not explain these very same facts.

Just as Wallace (1855) used Darwin's examples from the Galapagos to outline a process analogous to what is currently termed vicariance, he used his own field-based discoveries to show that such a process could lead to relationship, though not identity between the faunas of Australia and New Guinea:

> At the period when New Guinea and North Australia were united, it is probable that their physical features and climate were more similar, and that a considerable proportion of the species inhabiting each portion of the country were found over the whole. After the separation took place, we can easily understand how the climate of

both might be considerably modified, and this might perhaps lead to the extinction of certain species. During the period that has since elapsed, new species have been gradually introduced into each, but in each closely allied to the pre-existing species, many of which were at first common to the two countries. This process would evidently produce the present condition of the two faunas, in which there are many allied species—few identical. (Wallace, 1857, in Brooks, 1984, p. 161)

In this publication Wallace also related the changes in faunal content of New Guinea and Australia to a series of geological changes:

It is evident that, for the complete elucidation of the present state of the fauna of each island and each country, we require a knowledge of its geological history, its elevations and subsidences, and all the changes it has undergone since it last rose above the oceans ... a knowledge of the fauna and its relation to that of the neighboring countries will often throw great light upon the geology, and enable us to trace out with tolerable certainty its past history. (Wallace, 1857, in Brooks, 1984, p. 163)

Thus, Wallace recognized not only that geological events influence evolutionary patterns, but that we can use faunal histories to infer something about geological histories. Many research programs in paleobiogeography are dedicated to using biogeographic patterns in fossils to uncover something about geological history [e.g., see Scotese and McKerrow (1990), Fortey and Cocks (1992), and Lieberman and Eldredge (1996)]. This was a crucial insight and it provides more evidence that Wallace had incorporated the core of the idea that was to provide inspiration for the current vicariance biogeography movement.

4.3.2. The Darwinian Notebooks

Although Wallace was the first to tie the discipline of biogeography to evolution in a published work, it is clear that Darwin in the years between his voyage aboard HMS *Beagle*, and the publication of *On the Origin of Species* ... was thinking deeply about what the facts of geographic distribution could tell us about evolutionary divergence. In the notes he made toward the end of his trip around the world, the so-called ornithological notes, he recalled his observations on the Galapagos Islands, and recognized their potential significance:

When I see these islands [of the Galapagos Archipelago] in sight of each other and possessed of but a scanty stock of animals, tenanted by these birds but slightly differing in structure and filling the same place in Nature, I must suspect they are only varieties .... If there is the slightest foundation for these remarks, the zoology of archipelagoes will be worth examining; for such facts would undermine the stability of species. (Darwin, in Kottler, 1978, p. 281)

The birds Darwin was referring to were mockingbirds (Kottler, 1978; Sulloway, 1979). Although at this point he was not yet a believer in evolution by common descent, it was clear that he was swaying toward that view. Mayr (1976) in fact argued that it was the case of the mockingbirds of the genus *Mimus* from the Galapagos that convinced Darwin to become an evolutionist.

When Darwin returned from the voyage aboard the *Beagle*, he compiled a set of notebooks. In the first of these, begun in 1837, he inserted a note in 1838, reading "In July (of 1837) opened first note book on 'transmutation of species'. Had been greatly struck from about month of previous March on character of South American fossils, and species on Galapagos Archipelago. These facts origin (especially latter) of all my views" (Darwin, in Kottler, 1978, p. 280). The general validity of this statement can be shown by going back to the examples from Darwin (1839), where he cited the two South American ostrich species that showed exclusive spatial distributions, while the extant and extinct South American ungulate mammals had a mutually exclusive chronological relationship.

Originally, at least publicly, Darwin did not view these singular facts as necessitating a belief in evolution. However, in subsequent years, during the genesis of his many notebooks, it became clear to him that his initially disconnected observations on the flightless birds and the ungulates suggested that the processes leading to geographic differentiation between species must also have led to evolutionary changes through time (Sulloway, 1979; Browne, 1983). Here was a case in which Lyell's (1832) aphorism "As in space, so in time" might explain the origin of species. Interestingly, the fossil llamalike mammals (the macrochaenids) and modern llamas are no longer believed to be closely related (Gould, 1996), so while Darwin came up with the right answer to explain the evolution of life, some of the underlying data were inaccurate. Later his support of Lyell's aphorism was made even more explicit: "laws governing the succession of forms in past times being nearly the same with those governing at the present time the differences in the different areas" (Darwin, 1872, p. 384).

In the earlier sections of his notebooks, Darwin ascribed an important role to geographic isolation as leading to the origin of new species (Grinnell, 1974; Mayr, 1976, 1982; Kottler, 1978; Sulloway, 1979; Richardson, 1981). Geographic isolation would act to produce new species by facilitating their variation and making them reproductively isolated. He discussed how the isolation of populations of a species from one another could not be produced by distance alone but required some barrier as well, and, further, was related to the intrinsic ability of organisms to disperse (Kottler, 1978). Originally, he held that isolation would generally be produced by geological change, which would motivate the divergence and evolution of species. Such geological change included the elevation and subsidence of islands (Grinnell, 1974; Sulloway, 1979). Darwin (1909) described an example with islands, clearly meant to represent the Galapagos Islands, that illustrated this well. He posited that a group of islands might rise, then join, split, and become partially submerged. When they joined, floras and faunas would mix. When they sank they would become separated and their floras and faunas would become isolated. Eventually, when geological changes brought the islands back together again it would lead to variation and change in the floras and faunas owing to natural selection induced by interorganismic competition (Browne, 1983).

Thus, for Darwin, it was not the actual isolation of faunas and floras on different islands that was driving evolutionary divergence, but rather that divergence was produced by bringing species into contact with one another after they had been separated for long periods of time. Still, he was viewing the Earth as a dynamic, external agent engendering changes in species (Browne, 1983). The biological organisms were essentially passive, being driven by geological changes that were contingent [in the sense of Gould (1989)] on historical events. Thus, at least initially, Darwin held that there was a close association between global change and evolutionary and biogeographic change. Such a view implies that studying the history of life and its relationship to geologically driven changes would be an important way of studying evolution.

By contrast, later in his notebooks, around 1844, Darwin came to place less emphasis on isolation (Richardson, 1981), and actually came to see it as a potential barrier to change. As an example he cited the echidna, which seemed to him to be a very primitive mammal, but which was found as part of the isolated biota of Australia (Grinnell, 1974). His views on biogeography began to change such that he focused on the ability of species to disperse "as far as barriers, the means of transportal, and the preoccupation of the land by other species would permit" (Darwin in Browne, 1983, p. 197). This was very similar to what Lyell (1832) had said. Associated with this, there would be evolutionary divergence, with populations on the fringes of a species' distribution more likely to be exposed to struggles and competition with other species and thus more predisposed to change. This emphasis on naturally arising divergence and dispersal led Darwin to a concomitant deemphasis on geological change as an agent of evolution. Instead, he began to concentrate more on the internal qualities of species that lead to divergence— active, biological mechanisms independent of geological change. Thus, Darwin's newer formulation was more mechanistic, relying less on contingent, historical, geological events and more on intrinsic processes. The newer formulation is the one that dominates in *On the Origin of Species*...although even there he considers isolation an important contingency that can lead to evolutionary change, as will be discussed more fully below. Thus, through time Darwin came to disassociate global change and evolution. His later perspective implies that looking for a relationship between geological change and evolution in the fossil record would not be a fruitful way to analyze evolutionary patterns.

4.3.3. Darwinian Biogeography in *On the Origin of Species*...

Although Darwin's intellectual legacy and his contribution to the development of evolutionary theory have always been strongly tied to the discipline of biogeography, as evidenced by the opening sentence in the *Origin*, which I quoted in my Preface, it can be seen that his attitudes toward biogeography shifted throughout his career. Further, consideration of the

Darwinian notebooks is worthwhile, but their contents do not represent those of his writings on biogeography that were to influence his contemporaries as well as scientists of subsequent generations. Until fairly recently, little was known of them. His views on biogeography that were most influential were those presented in the *Origin*. In that book he had two chapters in which he discussed the geographical distributions of organisms and their bearing on the theory of evolution. He felt that although there were differences among the faunas of various regions, there was something that connected them. These regions did not represent separate centers of creations. Instead, there is "some deep organic bond, throughout space and time, over the same areas of land and water, independently of physical conditions. The naturalist must be dull who is not led to enquire what this bond is. The bond is simply inheritance" (Darwin, 1872, p. 341). Thus, it is clear that Darwin thought something, common descent, tied these faunas together, which was important. "Allied species, although now inhabiting the most distant points, have proceeded from a single area—the birthplace of their early progenitors" (Darwin, 1872, p. 370). He strikingly emphasized this in the case of the similarity, but not identity, between the faunas of South America and the Galapagos Islands.

An equally important question, if one were to deny that these separate faunas were created independently, was to determine how they might have diverged. Like de Candolle père, and Wallace, Darwin realized that "in considering the distribution of organic beings over the face of the globe, the first great fact which strikes us is that neither the similarity nor the dissimilarity of the inhabitants of various regions can be wholly accounted for by climatal and other physical conditions" (Darwin, 1872, p. 339). As I noted above, after returning from his voyage aboard the *Beagle*, Darwin came to rely on isolation as a means of producing evolutionary divergence, and this reliance is still partly in evidence in the *Origin*. For example, he wrote that, "Barriers of any kind, or obstacles to free migration, are related in a close and important manner to the differences between the productions of various regions" (Darwin, 1872, p. 340).

Darwin (1859, 1872) concluded that widely ranging species will spread out, beat out competitors, and seize on new lifestyles. The mere act of spreading out leads to encountering new conditions and divergence. Barriers will serve as a check to migration, thereby allowing natural selection to act:

> The dissimilarity of the inhabitants of different regions may be attributed to modification through variation and natural selection...the degrees of dissimilarity will depend on the migration of the more dominant forms of life from one region into another having been more or less effectually prevented, at periods more or less remote. (Darwin, 1872, p. 341–342)

Darwin, like Lyell (1832), believed that populations of a single species can become isolated owing to geographic and climatic changes, which can turn what was once a species with a single continuous range into a species comprised of several disjunct populations. Closely related species in disjunct areas could be produced by extrapolating the following sequence: continuous range, discontinuous ranges, and finally divergence. "The existence of closely allied...species in any two areas implies, on the theory of descent with modification, that the same parent-forms formerly inhabited both areas" (Darwin, 1872, p. 440).

Although Darwin wrote that isolation was important to his theory of evolution in both the notebooks and in the *Origin*, in the latter he posited that migration and isolation alone cannot effect evolutionary change. For example, if a species were to move to a new area along with its competitors, it would not diverge. In a response to Moritz Wagner, a proponent of isolation as the mechanism of speciation, he stated that "I can by no means agree that migration and isolation are necessary elements for the formation of new species...I believe that many perfectly defined species have been formed on strictly continuous areas" (Darwin, in Mayr, 1976, p. 120–121). He stated further (Darwin, 1872, p. 105) that "I can bring forward a considerable body of facts showing that within the same area, two varieties of the same animal may long remain distinct, from haunting different stations, from breeding at slightly different seasons, or from the individuals of each variety preferring to pair together." Evolutionary change within lineages comes when new types of organisms are contacted, and to a lesser extent, when new physical conditions are contacted (Darwin, 1872, p. 342, 357).

Thus, Darwin (1859, 1872) believed that it is not so much the act of isolation, but the exposure to a new selective milieu, that leads to divergence. "Although isolation is of great importance in the production of new species, on the whole I am inclined to believe that largeness of area is still more important" (Darwin, 1872, p. 107). He believed that in larger areas there would be more individuals and a better chance of favorable variations such that the conditions of life would become more complex. Further, he held that this would favor the origin and spread of new varieties (Mayr, 1976). The greater the competition that any population experienced, the greater its fitness. Finally, he believed (Darwin, 1859, 1872) that the fitness of a population was correlated with the size of the region it inhabited, such that the floras and faunas of small islands, or smaller island continents such as Australia, and the southern continents, would be generally less competitively fit than their counterparts in the larger northern continents (Sulloway, 1979). Wallace and the paleontologist William Diller Matthew supported this view that higher types emerged in the north to dominate elsewhere (Bowler, 1996). This viewpoint was unfortunately invalidly utilized as a biological argument for the supremacy of northern races of humans.

Darwin also believed that accidental transport or chance long-distance dispersal could isolate populations of species (Darwin, 1872, p. 344), and he performed many experiments on seeds to test the potential for long-distance dispersal. Reminiscent of the discussions by Lyell (1832) on the ability of species to disperse over long distances, this was important. Darwin believed that if most of the species "inhabiting one region are different from those of another region, though closely allied to them, it can be shown that migration from one region to the other has probably occurred at some former period"

(Darwin, 1872, p. 345). Thus, it is clear that Darwin favored migration and dispersal as generating biogeographic propinquity rather than vicariance, where broad ancestral distributions become divided up into more narrowly defined species ranges. Of course, the originally widely distributed ancestral species might have gotten its broad range by migration. The emphasis in Darwin (1859, 1872) would appear to be that evolutionary affinity and migration are tied together more fundamentally than evolutionary affinity and vicariance. "All the grand leading facts of geographical distribution are explicable on the theory of migration, together with subsequent modification and the multiplication of new forms" (Darwin, 1872, p. 382). This reliance on dispersal as the mechanism generating biogeographic patterns is similar to what von Buch (1825) outlined, and Darwin had read this work prior to the publication of the *Origin* (Kottler, 1978).

Darwin's views that chance dispersal contributed prominently to biogeographic patterns were partly based on his belief that the Earth's continents were static. For instance, he believed that there has been no "vast change in the position and extension of our continents" (Darwin, 1872, p. 346). Further, he did "not believe that it will ever be proved that within the recent period most of our continents which now stand quite separately have been continuously, or almost continuously, united with each other" (Darwin, 1872, p. 347). Of course, with the development of plate tectonics in the 20th century, this opinion proved to be incorrect.

4.3.4. The Role of Dispersal in the Darwin's and Wallace's Biogeography

Darwin's views on the role of dispersal in biogeography were very much in contrast with Wallace's early views. At least originally, Wallace was opposed to using long-distance, chance dispersal to explain the distribution of species. Wallace (1857) related biogeographic patterns to geological histories by arguing that if regions that currently share similar faunas are now separated, it is likely that these regions were formerly joined. He applied this explanation to his experiences with the faunas of the Malay Archipelago:

> The distribution of the animals of Aru and New Guinea (both in the Malay Archipelago) proves the close connection between these countries, it being evident that, where a considerable number of animals which have no means of passing from the one to the other are common to two countries, some former communication must have existed between them. A few such cases of community may indeed be explained by the various accidents by which animals may be transported from one country to another; but when the community is more general, there is no such easy way of accounting for it. (Wallace, 1857, in Fichman, 1977, p. 47)

Wallace (1860) reiterated this, arguing that even under favorable conditions, accidental transport is very unlikely (Fichman, 1977).

In this regard, Wallace was originally a strong supporter of the view outlined by the great botanist and friend of Darwin, Joseph Hooker. Hooker wrote a long treatise on the flora of New Zealand in which he posited that, "Similarities between the flora of New Zealand, Tasmania, and temperate South America were due to their being remnants of a flora that had once spread over a larger and more continuous tract of land than now exists in that ocean" (Hooker, 1853, in Fichman, 1977, p. 49). The idea propounded by Hooker (1853) that Wallace later adopted is equivalent to a vicariance view: similar floras and faunas were once part of areally more extensive regions that later fragmented. Widespread species would differentiate into more narrowly distributed species, but the species in the now separated regions would share close evolutionary affinity. Thus, evolutionary affinity among the species of separated regions would be evidence that the regions had once been joined. Hooker believed that it is geological and climatic factors that lead to evolutionary divergence owing to vicariance (Brown and Lomolino, 1998). Such geological and climatic factors could cause joined regions to separate.

Murray (1866) and Huxley (1870) took similar viewpoints. They also argued that different geographical provinces than those of the present day ones existed because organisms tended not to move except when geological conditions changed yet similar types of organisms span modern geographic barriers (Desmond, 1982; Bowler, 1996). Hooker, Murray, and Huxley all postulated a Pacific supercontinent that fragmented, though they disagreed about its precise location. Earlier, Forbes (1846) had invoked a large continent, now submerged in the Atlantic, which he called Atlantis, to explain patterns of similarity between Irish and Spanish faunas. Generally, those who favored this view were also more sympathetic to the existence of now submerged land bridges joining continents.

Eventually Wallace came to shift his views such that they became more aligned with those of Darwin. He (Wallace, 1863) came to believe, e.g., that the continents had moved little throughout Earth history. He still accepted the distinctness of Sclater's biogeographic regions, but he used this idea to argue that the existence of these regions was due to major features of the Earth's surface that had not undergone any relatively recent change, and that the oceans and the continents had not moved (Fichman, 1977). Later, he retreated even further from his vicariance perspective that Earth history changes were what structured the Earth's fauna. He, stated (Wallace, 1876), e.g., that dispersal alone could explain the biogeographic distributions of animals, though 4 years later (Wallace, 1880) he still stressed the role that sea-level fall must have played in joining what are now regions separated by water [see Michaux (1991) and Boer and Duffels (1996) for further discussion of the fact that Nelson and Platnick (1981) and Croizat (1982) mischaracterized Wallace when they treated him as a pure dispersalist, even late in his career].

Gulick (1888) also firmly believed in the role of dispersal as an agent of biogeographic change. In his view organisms tended to wander such that there was a law of migration of organisms, and that for geographic isolation to occur, they had to cross ecological barriers set by their preferred habitats (Kottler, 1978).

The belief that species and faunas were continually dispersing outward corresponded to the imperialistic metaphor prevalent in 19th-century Europe and Victorian England (Desmond, 1982; Bowler, 1996). At that time, there was aggressive expansionism from Europe out to the southern continents, so there were probably cultural reasons that led Darwin and later Wallace (though apparently not Hooker, Huxley, and Murray) to accept dispersal rather than vicariance as an explanation of the geographic distribution of species. Darwin was also not willing to accept the existence of land bridges, large sunken continents, or major changes in the configuration of the Earth's continents (Desmond, 1982). In the end, this can probably be attributed to his uniformitarian, actualistic approach (Mayr, 1982), which was derived from Lyell (1831, 1832). Both men believed in explaining geological events using modern processes and rates. The problem of conflating different types of uniformitarianism was nicely pointed out by Gould (1965).

With our modern understanding of geology and geophysics there now exists, in the framework of plate tectonics (discussed more fully in Chapter 6), a clear mechanism that can explain the patterns of distribution that Hooker, Huxley, and Murray uncovered. Regions were once joined and then they separated, so that there has been a dramatic change in the geometry of the Earth's plates. Thus, Darwin and Wallace were wrong about the immobility of continents, and part of the extensionist's arguments must hold sway. However, this does not mean that dispersal, or a process analogous to it, has never influenced biogeographic patterns. The relative contributions of vicariance and dispersal to paleobiogeographic patterns is discussed more fully in Chapters 5 and 6.

4.4. The Role of Isolation as a Mechanism of Speciation and Biogeographic Differentiation

As I noted earlier, although Darwin originally relied on isolation as an important process leading to speciation and biogeographic differentiation, he later changed his mind. Wallace's ideas followed a similar trajectory, in consonance with his changing views on the role of chance dispersal as an important process in biogeography. He originally held isolation to be an important process, but later came to believe that "geographical or local isolation is by no means essential to the differentiation of species, because the same result is brought about by the incipient species acquiring different habits or frequenting a different station" (Wallace, 1889, p. 150).

Moritz Wagner was a prominent natural historian who fought against Wallace's and particularly Darwin's rejection of isolation as the primary mechanism of speciation and thus biogeographic differentiation. To Wagner, isolation was the *sine qua non* of speciation (Mayr, 1976). Tied in with the requirement for isolation, Wagner (1868) also suggested that migration was necessary (Sulloway, 1979). "The formation of a genuine variety...will succeed in nature only when a few individuals can spatially segregate themselves for a long time from the other members of the species by transgressing the confining barriers of their range" (Wagner, 1889, from Mayr, 1976, p. 123).

The emphasis on isolation in evolution implicit in the works of such 19th-century natural historians as von Buch, Wagner, and, at least early in their careers, Darwin and Wallace, faded, largely due to the pervasive influence of a recanting on the part of the latter two and their followers. However, the seed that would allow a renewed appreciation for isolation had been sown. The likely reason that isolation never took hold originally as a primary explanation of evolutionary change is that it was not integrated into a discussion of mechanisms of evolutionary change. Rather, it was in the nascent stage of describing a pattern: the distribution of closely related species separated by geographic barriers.

The history of science is replete with such examples of valid pattern-based explanations that were originally rejected, to be accepted only later by the general scientific community when a mechanism accounting for such patterns is discovered (Gould, 1978). One excellent example of this is continental drift, the hypothesis that the Earth's continents have changed their position through time. Originally described based on patterns such as the geometry of continental margins and the distribution of fossil groups, it was roundly rejected by many geologists because no valid mechanism existed at the time to explain how the continents might have moved around. Later, however, with the descriptions of plate tectonics, the old hypothesis of continental drift, in new clothes, found universal acceptance in the geological community.

4.5. Conclusions

The one man who can be credited with almost single-handedly linking isolation and evolution by demonstrating the role of the former as a mechanism of speciation is Ernst Mayr. His years of arduous field work, hundreds of scientific papers, and numerous books spread the gospel of geographic speciation as a, perhaps the, mechanism of evolutionary change. In the next chapter I will discuss some of the different perspectives developed by Mayr and other 20th-century scientists and emphasize their deep significance for thinking about biogeography and paleobiogeography.

Chapter 5

Allopatric Speciation and Vicariance

5.1. Introduction

In evolutionary biology in the 20th century few intellectual figures stand out like Ernst Mayr. His contributions are so numerous that it would be impossible to present them in this book's limited format. However, chief among them was certainly his dogged devotion as a tireless advocate for the prevalence, in fact, the near universality, of the role of geographic isolation in speciation. He has numerous publications demonstrating the phenomenon of geographical speciation that predate his *Systematics and the Origin of Species* (Mayr, 1942), which is one of the three great documents of the evolutionary synthesis, but that book stands alone as a remarkably clear description of the phenomenon and its relevance to evolutionary biology.

Focusing on patterns of intraspecific differentiation, Mayr strongly endorsed von Buch's (1825) and Wagner's (1889) views on the prevalence of what we would now call allopatric speciation by demonstrating the important role of barriers in speciation (Mayr, 1942, p. 156). He recognized that hand in hand with the correlation between isolation and interspecific differentiation were patterns of geographic variation within species, and that "no fruitful discussion of the species problem is possible without a full understanding of the problems of geographic variation" (Mayr, 1942, p. 103).

Mayr (1942) also emphasized the importance of careful taxonomic studies by systematists who sample the entire geographic range of a species, and so capture geographic variation. By focusing on patterns of geographic differentiation within species, rather than just across species, he made it easier to conceptualize the intimate links among barriers, reproductive isolation, evolutionary divergence, and eventually speciation. Further, he provided numerous examples of anatomical characters that varied geographically to back up his contention. He sometimes treated sympatric speciation, the notion that populations of a species can diverge within the same region owing to ecological segregation, natural selection, and competition, without the intervention of geographic barriers, dismissively. He believed that

> ... it is now becoming more certain with every new investigation that species descend from groups of individuals which become separated from the other members of the species, through physical or biological barriers, and diverge during this period of isolation. (Mayr, 1942, p. 33)

Subsequent reviews and studies have largely tended to confirm this view (e.g., Mayr, 1982; Wiley and Mayden, 1985; Lynch, 1989; Funk and Brooks, 1990; Brooks and McLennan, 1991; Allmon, 1992).

Sympatric speciation, essentially the process of divergence that was strongly emphasized by Darwin (1859, 1872), appears to play a limited role in evolutionary divergence and speciation. That is not to say that no cases of sympatric speciation have been documented. Probably the most powerful examples of, and arguments for, sympatric speciation were put forward by Guy Bush, and the interested reader should see Bush (1969) for examples of possible sympatric speciation in action.

The phenomenon that Mayr identified as geographic or allopatric differentiation had been recognized earlier, as documented by Croizat *et al.* (1974), e.g., what Jordan (1908) called geminate species are species that are adjacent and closely related, but separated by a geographic barrier. However, this in no way takes away from the profound importance of Mayr's career and his work.

5.2. The Relevance of Allopatric Speciation to Historical Biogeography

The significance of allopatric speciation to problems in historical biogeography was quickly recognized by early supporters of what has come to be called the vicariance school of biogeography. Croizat *et al.* (1974) viewed allopatric species as equivalent to vicariant species, which would form after barriers preventing gene flow developed within a formerly continuous population such that the populations became separated. This corresponds to Croizat's (1964) term "immobilism" and Brundin's (1988) "vicariant form making." Conceivably, two closely related species that were separated by a geographic barrier might be produced by the dispersal of a population of an ancestral species over a geographic barrier, with that dispersing population subsequently differentiating. However, this explanation was treated as having limited generality and utility for problems in biogeography, as will be discussed more fully below.

Croizat *et al.* (1974) also argued that cases of apparent sympatry, where two closely related species had identical ranges, were likely to reflect original vicariant, allopatric differentiation, with subsequent dispersal by one or both of the new species into the range of the close relative. Thus, in their work there is an emphasis on the primacy of allopatric speciation.

However, the definition of allopatric speciation utilized by Croizat *et al.* (1974) and later Platnick and Nelson (1978) is insufficiently precise. Actually,

> ...allopatric speciation is an umbrella term for a spectrum of modes that involve complete geographic separation of two or more subpopulations of a species during its evolution into two or more daughter species (one of which may be the ancestor itself). (Wiley and Mayden, 1985, p. 600)

There are at least three distinct modes of allopatric speciation: vicariance, peripheral isolation, and multiple peripheral isolates speciation (Wiley and Mayden, 1985; Funk and Brooks, 1990; Brooks and McLennan, 1991). There is also the related alloparapatric/parapatric model of speciation. Vicariance speciation as defined by Wiley and Mayden (1985), Wiley (1988*a*), Funk and Brooks (1990), and Brooks and McLennan (1991) is essentially concordant with Croizat *et al.*'s (1974) or Platnick and Nelson's (1978) definition of allopatric speciation. In vicariance speciation, what is originally a single widespread species is fragmented by geographic barriers which leads to differentiation and then to speciation, with the creation of new daughter species. Under this model, the total sum of the ranges of the daughter species will be equivalent to the range of the ancestral species (Wiley and Mayden, 1985; Brooks and McLennan, 1991) (Fig. 6). Brooks and McLennan (1991) characterized this mode of speciation as largely passive from an organismic perspective, because it generally proceeds owing to geological or climatic changes that create barriers.

In allopatric speciation by peripheral isolation, an ancestral species colonizes a small new area by passing over a barrier and subsequently differentiates (Wiley and Mayden, 1985; Brooks and McLennan, 1991). In this mode, the species in the peripheral isolate often shows a greater degree of morphological transformation than the change seen in the parent species in the ancestral range. This pattern of differential change in the pair of allopatric species has been termed Hennig's (1966) deviation rule because one species shows greater deviation in its morphology. In contrast to the vicariance mode of allopatric speciation, this mode was characterized by Brooks and McLennan (1991) as active because it relied on the movement of organisms into new regions over barriers. In multiple peripheral isolates speciation, numerous small populations of the species are isolated from one another such that they all differentiate and diverge at roughly the same time.

Finally, alloparapatric and parapatric modes invoke similar types of processes. In the former, populations of a species began to differentiate after the emergence of barriers and come into secondary contact after they have differentiated but while they are still reproductively viable. Speciation then occurs owing to interactions between the differentiated populations. In the latter, there is a continual narrow zone of overlap or sympatry between two populations of a single ancestral species. Differentiation develops and speciation eventually occurs in spite of this primary overlap in parts of the ranges of these populations (Wiley and Mayden, 1985; Brooks and McLennan,

1991). The primary distinction that I want to emphasize among these different modes is that, from the perspective of the organism, geographic speciation is either passive or active. Geology and climate are either the primary and secondary motivating forces of evolutionary change (they cause the initial isolation and the isolation causes the divergence) or they are just the secondary motivating forces (they do not cause the initial isolation but the isolation still causes the divergence).

Why was the recognition of the prevalence of allopatric speciation important for the development of the field of biogeography? First, it placed primary emphasis on geographic barriers in motivating speciation, and conceivably geographic barriers for one group would in many, though not all cases, be geographic barriers for other groups. Second, the establishment of such geographic barriers would frequently be due to geological or climatic change, especially in the vicariant mode of allopatric speciation. This meant that episodes of geological change potentially set the stage for speciation in a large number of groups that occurred in the affected area (Funk and Brooks, 1990). Thus, a reliance on allopatric speciation offered a potentially integrative framework with which to approach the problem of the relationship between geographic distribution and evolution in groups of organisms. In this case, separate explanations for biogeographic patterns might not be necessary for each group of organisms that occurred in a region. Instead, these patterns might have been governed by a single event. Thus, allopatric speciation, in the vicariant mode, would be expected to produce the fundamental pattern of historical biogeography (see Fig. 2). The fact that most speciation events occurred via allopatry and that vicariant speciation is the most frequent mode of allopatric speciation (discussed more fully below), means that there should be a strong historical biogeographic signal in any biota occupying a particular region.

Allopatric speciation, particularly by the vicariant mode, has a deeper significance for evolutionary biology that relates to contingency, a phenomenon whose importance to the field has been thoroughly demonstrated by Gould (1989). If vicariance is the primary way that speciation occurs it means that speciation events across an entire biota will be structured by episodes of geological and/or climatic change. These changes are by their nature highly contingent on historical processes, i.e., they involve sequences of historical events, such as the separation of continents. It also implies a close correspondence between the history of Earth and the history of life. [As Croizat (1958, 1964) suggested, the Earth and its biota would have coevolved] and that speciation and macroevolutionary change cannot be explained primarily by deterministic biological processes such as competition and natural selection. Rather, speciation events will be initiated by geological events. What happens to populations after they become isolated and as they began to differentiate still may be governed by natural selection, competition, and other deterministic biological processes explicable by the principles of population genetics; however, geological change becomes the initiating factor. The relationship between geological change and evolution also implies an important role for paleontology within evolutionary biology because the fossil record offers excellent potential for studying how evolution is associated with geological change over long periods of time. If vicariance is important, we cannot understand the evolution of the modern biota without looking at past geological events.

5.3. Comparing Sympatric and Allopatric Speciation

In contrast to allopatric speciation, sympatric speciation would be unlikely to produce historical biogeographic patterns. The mechanisms driving sympatric speciation relate to ecological differentiation, competition, and natural selection. As every species tends to have its own unique ecological characteristics, the ecological factors that might act to produce sympatric speciation in one species are unlikely to do so in another, the arguments of Endler (1982) notwithstanding. Further, if we imagine that hosts of species were produced by sympatric speciation, these species would all have the same range! Such a pattern is clearly contradicted by the grand scale in the history of life as reflected in both the modern and fossil biota. We must remember, that the fundamental pattern historical biogeography needs to explain is that different species occur in different areas—the notion that Nelson (1978) expressed as Buffon's law—and this pattern is incompatible with pervasive sympatric speciation. Further, all available analyses [reviewed in Brooks and McLennan (1991)] suggest that sympatric speciation occurs only rarely, so although not impossible, it offers little prospect for integration into an historical biogeographic framework. Hence, this variety of speciation warrants little attention in this study.

5.4. Why Speciation Is Important in Biogeography?

Thus far I have developed the notion that allopatry is the predominant way in which speciation occurs, and identified Ernst Mayr as the scientist who led the push to get this accepted. The reason a discussion of modes of speciation is important to paleobiogeography is manifold, and relates to themes I discussed in Chapters 2 and 3. Many theoretical perspectives have come to suggest that the primary way that macroevolution occurs is by speciation. In short, the history of life at the large scale of clades of monophyletic taxa is best explained by summing a great number of speciation events. These theoretical perspectives emerge from three interconnected areas in the philosophy of biology and macroevolutionary theory: a revised ontology of the nature of species [*sensu* Eldredge (1985b)], the theory of punctuated equilibrium [*sensu* Eldredge and Gould (1972) and Gould and Eldredge (1977)], and the distinction between anagenesis and cladogenesis.

If we accept that the history of life is best viewed as a set of speciation events, then obviously the mode by which speciation occurs becomes a vitally important point for evolutionary biologists. Moreover, if speciation is predominantly allopatric, as I have argued, then there is a close relationship between geographic distribution and speciation. It is not then a great leap to conclude that geographic distribution and the history of life are fundamentally linked. Indeed, it was the geographical distribution of closely related species that suggested to Wallace and Darwin that life had evolved.

In terms of the three interconnected areas in the philosophy of biology and macroevolutionary theory, the first, the revised ontology of species, refers to the growing contention that species have a real existence in nature. They are individuals as discussed by Ghiselin (1974) and Hull (1978, 1980) and more than just a simple sum of their parts. Further, they are more-making entities or replicators [see, e.g., Eldredge (1979, 1985a, 1989a), Vrba (1980), Gould (1982), Vrba and Eldredge (1984)], giving rise to more entities of like kind—in this case, species. This emphasis on species was presaged by Dobzhansky (1937), one of the architects of the synthesis, who recognized that the two major features that any theory of evolution must explain are diversity and discontinuity, with the latter arising from the existence of species.

The second of the three interconnected areas, punctuated equilibrium (punct. eq.) (Eldredge and Gould, 1972), is grounded in the tenet that species are stable morphologically throughout most of their existence. If we accept punct. eq., the bulk of morphological and evolutionary change in species lineages is concentrated at speciation. Further, punct. eq. posits that speciation occurs over a relatively short time span, 5,000 to 50,000 years, at least compared to the total duration of the species, which is usually on the order of millions of years (Eldredge and Gould, 1972; Gould and Eldredge, 1977). In addition, in punct. eq. speciation occurs in narrow isolated populations via allopatry and does not involve anagenetic transformation of one entire species into another without the generation of new diversity. Instead, one species splits into two or more, and new diversity is generated. Eldredge (1979) referred to the outlook that emphasized cladogenetic speciation as the taxic view of evolutionary change, which he distinguished from the transformational view that relied on anagenesis. If we accept punct. eq. and the revised ontology of species, the existence of species not only explains Dobzhansky's (1937) discontinuity, but speciation is the way in which diversity is generated.

Putting this information together provides a powerful argument that the history of life is best viewed in the context of speciation events, which, in turn, are best viewed in terms of the geographic distribution of these species. It would be hard to construct a better argument for the belief that biogeography is an important discipline within the field of evolutionary biology.

5.5. Why Speciation Is Especially Important in Paleobiogeography?

In Chapter 3, I considered in detail the quality of the fossil record as well as the limits of temporal resolution in analyzing that record. The fossil record is the filter through which all paleontological data must be viewed, and the strengths and weaknesses of paleontology are correlated with the nature of that filter. Further, when we seek to reconstruct biogeographic patterns in the fossil record, we have to view them through the same filter. If we consider the nature of the speciation process that I outlined above, we see that my exposition relied heavily on allopatric speciation and punct. eq. to account for speciation. If we accept punct. eq. then unless our limits of temporal resolution in the fossil record exceed 5000–50,000 years in acuity, actual population differentiation leading to speciation will not be visible there. In short, biogeographic differentiation occurring at hierarchical levels lower then the evolution of species within clades will never be accessible to paleontologists. The paleobiogeographer is thus restricted to the study of paleobiogeographic patterns related to the differentiation of clades of species. This perspective does not mean that analyses of patterns at lower hierarchical levels have no significance. Quite the contrary! They are vitally important. But just as paleobiogeographers will be unable to study biogeographic patterns at lower hierarchical levels and finer temporal scales, biogeographers will be unable to study anything comparable to paleobiogeographic patterns in the modern biota. Nor, because of the existence of hierarchies, will paleobiogeographers or biogeographers be able to extrapolate results from one temporal scale to another. Both groups of biogeographers—those working with the modern biota and those analyzing the fossil record—have the potential for unique contributions to the study of biogeography.

5.6. The Relationship among Allopatric Speciation, Punctuated Equilibrium, and Dispersal

To wind up this chapter, I want to discuss what association, if any, is implied between geographic speciation and dispersal in punct. eq. In Chapter 6, I will discuss the biogeographic phenomenon of dispersal much more thoroughly. As outlined earlier, evolutionary biologists have determined that there are basically two ways that the isolation stage of allopatric speciation can be initiated: passive, geologically driven vicariance, and active, organismally driven peripheral isolation. Both, approaches emphasize the role that geographic isolation plays in leading to speciation, but they differ in explaining how such isolation comes about. Which of these two types of speciation, the active or the passive, is most emphasized by punct. eq.? As originally described by Eldredge and Gould (1972), punct. eq. treated the peripheral isolates mode of allopatric speciation described by Mayr (1963) as the primary mechanism of speciation. That is, it initially relied on active, organism-initiated dispersal as the primary mode of geographic isolation.

Since the publication of Eldredge and Gould (1972), studies have been conducted by macroevolutionists working with the modern biota in which they have attempted to quantify the relative prevalence of different modes of speciation, chief among them being Wiley and Mayden (1985), Lynch (1989), and Brooks and McLennan (1991). These studies found that the passive, vicariant mode of speciation was more common than the active (from the organism's perspective) peripheral isolates mode, though each of these occurred much more frequently than sympatric speciation. If we accept these studies, there clearly is some role for dispersal by organisms in the speciation process. These studies also suggest that when we think about the relationship between allopatric speciation and punct. eq., perhaps we need not always require active dispersal of organisms to be tied up with the speciation process. Instead, at least much of the time, perhaps the factors that initiate isolation are geological or climatic, with organismal dispersal over preexisting barriers playing a less important role.

Salient analytical and theoretical studies here are those conducted by paleobiologists who have looked at speciation in the context of macroevolution in the fossil record. Among them Vrba (1985, 1992, 1996) demonstrated that climate change plays a very important role in initiating speciation in several groups of animals simultaneously. She documented this in connection with her elucidation of the Turnover-Pulse hypothesis. In this hypothesis, climate change, especially change associated with the degradation and fragmentation of the preferred habitat of organisms (Vrba concentrated on mammals from the tropics of South America and Africa) serves to fragment species into isolated populations which then tend to either speciate or go extinct. Several groups of tropical mammals showed concurrent increases in speciation and extinction rates during times of climate change in the Pliocene and Pleistocene. These changes involved cooling and drying in the tropics, which affected the tropical rain forests that they inhabited and depended on. The type of speciation engendered by these changes is clearly allopatric in the vicariant mode. In this case, climatic change initiated isolation of populations so the mode of allopatric speciation she emphasized is different from that emphasized by Eldredge and Gould (1972).

Now how does this relate to punct. eq.? First, Vrba is a supporter of punct. eq. [see, e.g., Vrba (1980, 1989)] so her work was clearly not intended as an attack on the theory. Rather, she emphasized a different mode of allopatric speciation to produce concurrent episodes of speciation in several groups during times of climatic change. Further, the Turnover-Pulse hypothesis has been strongly endorsed and supported by one of the original architects of punct. eq. [see, e.g., Eldredge (1989a, 1995)]. (This is not to say that Gould opposes the Turnover-Pulse; he definitely does not. Rather, he simply has fewer discussions of the topic in his writings than Eldredge does.) Each of these facts suggests that punct. eq. is compatible with allopatric speciation in the vicariant mode. In other words, active dispersal by organisms is no longer a necessary part of allopatric speciation in punct. eq., and punct. eq. is compatible with either mode of allopatric speciation. Dispersal by organisms may initiate the isolation that drives speciation events in the fossil record via punct. eq., or it may not. More studies need to address the extent to which dispersal is, or is not, tied up with the process of allopatric speciation to confirm how punct. eq. in the fossil record should best be viewed.

5.7. Conclusions

In conclusion, the arguments in this chapter can be divided into the following key ideas: the primary way in which macroevolution occurs is by speciation; macroevolution follows a punct. eq. pattern in the fossil record; allopatric speciation is the predominant way that speciation occurs; allopatric speciation is generally, though not always, initiated by geological or climatic causes.

Chapter 6

Vicariance, Dispersal, and Plate Tectonics

6.1. Introduction

Probably the greatest debate in the history of biogeography concerns the relative extent to which dispersal, as opposed to vicariance, can best explain the evolution and distribution of organisms. This is, in fact, a debate that even occupied many natural historians who did not believe in evolution. The different views of such scientists as Buffon, Linnaeus, de Candolle père, Hooker, Wallace, and Darwin were discussed in Chapter 4. What is relevant here, however, is how, in the modern scientific framework of paleobiogeography, we should view both vicariance and dispersal.

The significance of vicariance, as discussed in the last chapter, is twofold. First, it means that potentially many different types of organisms can be affected by large-scale geological events, which would produce congruent episodes of speciation. Secondly, episodes of geological differentiation of regions would correspond to episodes of evolutionary differentiation of biotas, and there would be a close coupling between the history of Earth and the history of life (Fig. 6). By contrast, dispersal has often been cast in a very different light, and it has been suggested that it does not produce congruent biogeographic patterns. This chapter discusses the similarities and differences that emerge between vicariance and dispersal and the relevance of dispersal as an historical biogeographic phenomenon. What has interfered with our understanding of the role of dispersal in biogeography is the fact that the word has been used in connection with many different processes that actually

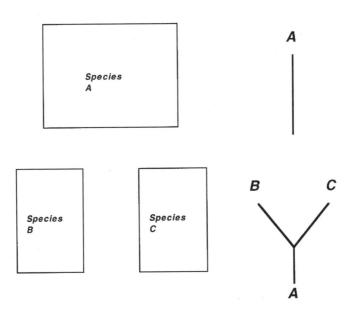

FIGURE 6. Vicariance in a phylogenetic perspective: (Top) a single species A distributed across an area; (Bottom) the area fragments in two, vicariance occurs, and a new species is found in each area. B and C are derived from A, and their combined ranges are equal to the range of A.

connote very different things. This chapter will describe these various processes and their relevance to the different subdisciplines of biogeography.

6.2. Dispersion and Dispersal

That most organisms move around during at least some stage of their life cycle is indisputable. Lyell (1832) and Darwin (1859) documented the ability of numerous organisms to travel great distances. Other important works conjoining evolution, distribution, and dispersal include Willis and Yule (1922), Darlington (1959), and Mayr (1963). The relationship between dispersal and certain types of allopatric speciation, which was discussed in Chapter 5 [see Wiley and Mayden (1985) and Brooks and McLennan (1991)], was codified by Hennig (1966) as the progression rule. Recent comprehensive surveys of dispersal that contain many interesting examples and an extensive literature were presented by Huntley and Webb (1989) and Brown and Lomolino (1998). The impressive evidence for dispersal makes it a cornerstone of island biogeography, one of the major theoretical approaches in the field (MacArthur and Wilson, 1967).

It has often been suggested that if dispersal were extremely pervasive there would be several consequences for the field of historical biogeography, all of them bad. This is because if organisms are always moving around then one might first predict that there would be little consistency between where a species arises from its ancestors and where it gives rise to new descendants, which would tend to produce incoherent historical biogeographic patterns within clades. Further, if dispersal were so rampant that all groups might be expected to be moving around in different directions, it would tend to produce incongruent historical biogeographic patterns across clades. Of course, neither of these notions is true. Rather, the significance of dispersal for historical biogeography depends on our theoretical perspectives concerning the relationship between microevolution and macroevolution as well as on the type of dispersal that is being invoked.

Many of the examples discussed by Lyell (1832), Darwin (1859), Huntley and Webb (1989), and Brown and Lomolino (1998) involve the movement of individual organisms or populations. Often times, there are associated concomitant shifts in entire species ranges, and these events occur on timescales from days to millennia. However, the events discussed by these authors do not involve speciation events; i.e., they are not macroevolutionary in scope (Funk and Brooks, 1990), and they may or may not involve microevolutionary change.

A term that has been used in the literature for the process that produces these types of biogeographic patterns is dispersion (Platnick, 1976). Although an overprofusion of terminology can be bad, much of the debate about the significance of dispersal for the field of biogeography may be about language, with different authors meaning different things when they talk about dispersal. Thus, I will henceforth use the term dispersion to connote such types of organismal and populational movement, associated possibly with microevolutionary change, but not with macroevolutionary change. Owing to the nature of the fossil record, paleobiogeographers will usually not be able to see dispersion.

6.3. Traditional Dispersal

Dispersal has generally been treated by phylogenetic or cladistic biogeographers as having only a small role in generating historical biogeographic patterns. To such biogeographers, who emphasize vicariance, traditional dispersal is treated as an expansion of range by a species over a barrier with essentially coeval diversification (e.g., Humphries and Parenti, 1986; Humphries *et al.*, 1988). It has also been referred to as the "second type of dispersal" by Platnick and Nelson (1978, p. 2). Based on this definition, this style of dispersal, unlike dispersion, has implications for macroevolution and thus historical biogeographic patterns.

Biogeographers and paleobiogeographers encounter several problems when trying to integrate this type of dispersal with historical biogeographic patterns. (Because the meaning for this type of "dispersal" seems well established in the technical literature, whenever it is invoked subsequently, I will refer to it as traditional dispersal.) First, such an event has traditionally been viewed as unlikely, or at least there are no substantiated examples of it occurring with the living biota (dispersal with the creation of a new species). Further, it would only occur in a single taxon and not be related to an episode of geological change, but rather be associated with specific ecological characteristics of the organisms within a species:

> The only conclusive evidence for chance [traditional] dispersal may be the demonstration that a given distribution is unique and unparalleled by that of any other living organism (Croizat *et al.*, 1974, p. 266) ... at present we are unable to identify a general causal principle of dispersal; we imagine that the causes of dispersal are as numerous as the species that have dispersed. (Croizat *et al.*, 1974, p. 269)

When traditional dispersal occurs, biogeographic histories of different groups of organisms are not expected to coincide, and thus it will not produce the element of congruence that is of such import to historical biogeographers (Croizat *et al.*, 1974; Rosen, 1978, 1979; Brooks *et al.*, 1981; Wiley, 1981, 1988*a,b*; Kluge, 1988; Brooks and McLennan, 1991).

Another problem with traditional dispersal as an historical biogeographic explanation is that it has been used in the past in a very *ad hoc* way. Anytime there was a seemingly anomalous distribution it was explained away by invoking one or more traditional dispersal events, to account for the biogeographic patterns in any group (Platnick and Nelson, 1978; Rosen, 1978; Nelson and Platnick, 1981; Wiley, 1981; Brundin, 1988; Kluge, 1988). One problem with such explanations is that they are rarely testable. Owing to the significant problems associated with invoking traditional dispersal the practice has been rightly criticized by vicariance biogeographers. Further, the existence of biogeographic congruence across various groups of taxa, and the presence of coherent biogeographic histories within individual clades, implies that traditional dispersal is not pervasive. Anytime congruence and coherence are recovered, the importance of traditional dispersal is diminished.

Although dispersion and traditional dispersal are the most frequently invoked types of "dispersal," they may not be the only kinds that historical biogeographers and paleobiogeographers need to be concerned with. In particular, I hope to demonstrate that there are cases in which "dispersal," but not traditional dispersal, can produce congruence across different clades of organisms. First, I want to discuss how it can happen. Then I will give a theoretical overview of its significance, propose a term for this type of dispersal, elucidate the original way in which it was addressed in a phylogenetic framework, discuss its significance to paleobiogeography, and then go back to some of the earlier literature where it was treated, albeit sometimes in a different context.

6.4. Defining a Different Type of Dispersal: Geodispersal

The recipe for producing congruent episodes of "dispersal" or range expansion is simple. There simply needs to be a change in the structure of the various areas clades occupy. Principally, a geographic barrier, either geological or climatic, has to fall such that several taxa can simultaneously, or almost simultaneously, expand their ranges via dispersal. Then, in many clades, the distribution of species will go from present in area A to present in areas A and B, although in actuality A and B have now been combined into one larger area (Fig. 7). This will be a congruent event, but at this point all that has occurred is a change in geographic distribution, a microevolutionary event: In other words, there are no macroevolutionary events: no cladogenesis and no speciation. Thus, this phenomenon is not immediately relevant to historical paleobiogeographers even though it is a congruent biogeographic event. For this type of "dispersal" to have relevance to historical paleobiogeography a second ingredient is needed. What is required is that subsequently the fallen

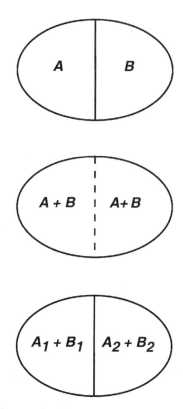

FIGURE 7. A hypothetical example of congruent range expansion or geodispersal: (Top) A continent bisected by a geographic barrier; prior to the emergence of this barrier there was a single ancestral species distributed across the entire continent. When the barrier emerged, populations became isolated on either side of it, differentiated, and then speciated via vicariance such that there were two closely related species separated by a geographic barrier. (Middle) At a later time, the geographic barrier was removed such that both species A and B could expand their range into new areas. (Bottom) Later, a geographic barrier reemerges. Now populations of both species A and species B are isolated on either side of the barrier. They differentiate and eventually speciate such that there are now a total of four species, two on each side of the barrier.

barrier, or perhaps another one within the newly expanded ranges, rises again producing allopatric speciation and vicariance in several taxa.

I have defined this type of "dispersal" as geodispersal (Lieberman and Eldredge, 1996; Lieberman, 1997). My original example involved a set of marine animal groups, trilobites, extinct relatives of horseshoe crabs and spiders, living in eastern North America, the Canadian Arctic, Europe, northern Africa and several other regions scattered across the globe 380 million years ago, during what is called the middle part of the Devonian period. At that time the geometry of the Earth's plates was very different than it is today (Fig. 8).

Plates were coming together and colliding, particularly in eastern North America, and these collision events produced the Acadian Orogeny or mountain-building event, the forerunner of the one that elevated the Appalachian mountain chain. There were also several major episodes of sea-level rise and fall that would have left the continents alternately inundated with water and thus available habitat for marine organisms such as trilobites, or largely dry and unavailable. There was strong evidence in these clades that

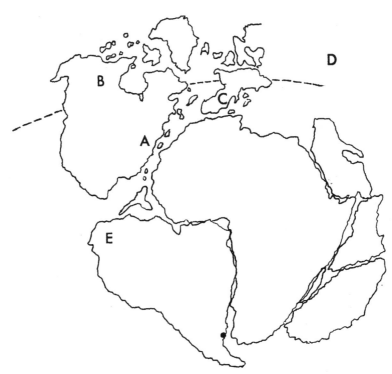

FIGURE 8. A paleogeographic reconstruction of the Middle Devonian. The dashed line is the paleoequator; the dot is the paleopole. A = eastern North America, B = the Canadian Arctic, C = Armorica, D = Kazakhstan, and E = northern South America. The precise orientation and geometry of Kazakhstan is poorly known; therefore, only its relative position is shown. Figure from Lieberman and Eldredge (1996), used with permission.

different groups showed congruent patterns of vicariance, but there was also evidence for congruent episodes of range expansion replicated across several of them. These episodes appeared to be related to geological events such as sea-level rise or continental collision, which would change the structure of the areas that organisms occupied, principally by merging two separate areas into a single larger one.

As these episodes of range expansion occurred in several clades simultaneously, I wanted to distinguish them from traditional dispersal, which involves dispersal over barriers, so I referred to this type of "dispersal" as geodispersal, to connote the close association between geological events and patterns of range expansion (Lieberman and Eldredge, 1996, p. 67). I then extended the use of the term geodispersal, so that it is defined herein as "the expansion of the range of a group of species due to the elimination of some topographic or climatic barrier" (Lieberman, 1997, p. 1039) followed by the emergence of a new barrier which produces subsequent vicariance.

6.5. Historical Framework on the Concept of Geodispersal

Although we (Lieberman and Eldredge, 1996; Lieberman, 1997) were the first to publish work that tried to rigorously integrate geodispersal into a geological and phylogenetic framework, the idea that there is such a phenomenon as geodispersal had been accepted by many authors and has a long intellectual pedigree. The kernels of the idea of geodispersal are found in the works of many early natural historians. The criteria I am using to identify early proponents of this process is whether or not they suggested that geographic barriers can fall, allowing animals and plants to expand their range, and then later rise, isolating animals and plants. For this reason, Linnaeus' ideas on biogeography [discussed in Kinch (1980), Browne (1983), and Brown and Lomolino (1998)] and ideas that invoked the movement (dispersion) of all animals from Noah's Ark after the Deluge are not treated as equivalent to geodispersal. Similarly, those who argued prior to the formulation of plate tectonics in the 1960s that land bridges existed will not be considered early advocates of geodispersal.

Probably the first scientist to suggest a process akin to geodispersal was Lyell (1832). He recognized that even the most prominent geographic barriers separating distinct regions could be effaced, and he argued that changes in the physical geography of the Earth can promote or retard the migrations of species (Lyell, 1832, p. 160). He also believed (Lyell, 1832, p. 169) that major climatic changes would cause the movement (geodispersal) of entire floras and faunas [also see Browne (1983)]. Lyell's ideas on this subject were part of his uniformitarian approach, through which he accepted that processes and events cycle continually through time.

Darwin, a great admirer of Lyell, subscribed to some of the latter's ideas about mechanisms controlling the distribution of animals. Specifically, he too believed that barriers could fall, joining biotas, and later rise, isolating them. In his notebooks Darwin described a specific example of islands joining and then splitting, and he argued that this might have led to the observed differences among living rhino species in Asia. For Darwin, joining and separating islands would have caused their component faunas to intermingle and then later separate. "Species formed by subsidence. Java & Sumatra. Rhinoceros. Elevate & join keep distinct. Two species made elevation & subsidence continually forming species" (Darwin, 1837–1838, in Barrett *et al.*, 1987, p. 191). The isolation of islands would result in isolated faunas, leading to evolutionary divergence.

However, Darwin also believed that evolutionary changes occurred when species were brought together because this would be greater selection pressure (Browne, 1983). "Springing up more likely to ⟨M⟩ have different species than those sinking, because arrival of any one plant might make conditions in any one island different" (Darwin, 1838, Notebook C, 25E, in Barrett *et al.*, 1987, p. 245). Thus, Darwin's views are not exactly analogous to the definition of geodispersal given herein, for he did not rely on allopatric differentiation, vicariance, and isolation as the primary factors motivating macroevolution. Instead, he believed differentiation would also occur during what I term the geodispersal phase.

Darwin (1837–1838, Notebook B), again probably based on his reading of Lyell (1832), also suggested that changing climates could induce large-scale migration of organisms, with concomitant diversification of taxa. He thus relied on both geological and climatic changes as forces motivating geodispersal and evolutionary differentiation. For example, he remarked, "Speculate on multiplication of species by travelling of climates & the backward & forward introduction of species" (Darwin, Notebook B, 202, 1837–1838, in Barrett *et al.*, 1987, p. 222).

In spite of these examples, or perhaps because of their rarity in Darwin's extensive notebook writings, it would be a mistake to argue that a process akin to geodispersal was a linchpin or even an important part of his thoughts on transmutation and biogeographic patterns. Careful combing of his notebooks reveal but two examples of anything analogous to geodispersal as defined herein. By contrast, he put great stock in both traditional dispersal and land bridges (e.g., Darwin, 1837–1838, Notebook B, 221, 223, in Barrett *et al.*, p. 226; Darwin, 1838, Notebook C, 25E, in Barrett *et al.*, p. 246; Darwin, 1838, Notebook D, 74E, in Barrett *et al.*, 1987, p. 357) as a means of explaining biogeographic patterns.

Wallace (1860) opposed traditional dispersal as a general explanation of biogeographic patterns and suggested that faunal similarities were likely to reflect what he referred to as strong links. These are geological connections. Although he later changed his views on the prevalence of traditional dispersal (e.g., Wallace, 1876), largely at the urging of Darwin (Fichman, 1977), early on he would likely have been sympathetic to a process such as geodispersal. Even later on (Wallace, 1876), where he emphasized the importance of traditional dispersal he described phenomena, such as the mass movement of organisms from the northern to the southern continents (see Bowler, 1996), which resemble geodispersal. For example, he suggested that into the southern continents "flowed successive waves of life, as they each in turn became temporarily united with some part of the northern land" (Wallace, 1876, p. 155). Further, he believed that South America was formed by the welding together of several smaller regions (Wallace, 1876, p. 27). Wallace (1876) conjoined his waves of emigration to falling geographical barriers and, in this respect, his ideas bear some resemblance to geodispersal, for falling barriers would allow taxa to expand their ranges.

Huxley (1870) also appears to have been an early proponent of a phenomenon akin to geodispersal. "For Huxley, the abrupt appearance of new types in the fossil record was the result of geological changes which permitted the invasion of new territory by forms already evolved in unknown areas" (Bowler, 1996, p. 392). Huxley argued that some of the herbivorous mammals of Africa and India were derived from a European fauna "which had migrated south after a sea barrier had been removed and had flourished in their new home" (Huxley, 1870, p. 374, from Bowler, 1996, p. 392). Subsequently these faunas were isolated by the formation of other geographical barriers such as the Himalayas, the Sahara Desert, and the Red Sea. His description of this phenomenon corresponds to the hallmarks of geodispersal: a pulse of range expansion following the fall of a barrier, and then the subsequent potential for vicariance when the same, or another barrier, should emerge.

There are several other examples of natural historians of the late-19th and early-20th centuries who subscribed to similar views including Beddard (1895, p. 227) "a continuous efflux of waves of life spreading out from the place of origin push further away the races which have the start" and Lydekker (1896). The concepts that Wallace (1876), Beddard (1895), and Lydekker (1896) discussed differ in an important respect from geodispersal because their ideas relied heavily on the role of competition as a force driving the movement of organisms. These ideas on competition are no longer generally accepted by evolutionary biologists, and the idea of active competitive urges leading to the outward expansion of organisms finds more analogies with European imperialism than with the history of life.

However, not all natural historians of that period relied on competition as the primary force driving organismal range expansion. Just as Wallace (1876) conjoined animal migrations to falling geographic barriers, Wortman (1903) invoked physical, i.e., climatic, changes as initiating factors for migration (see Bowler, 1996, p. 403). Matthew (1915, 1939) also postulated that climatic changes, along with competition, played a role in inducing large-scale migration of organisms, in this case from northern to southern continents. One famous example of large-scale movement well known to Darwin, Wallace, and Matthew was the Great American Interchange, the event that transpired when the Isthmus of Panama came into contact with South America as a result of plate tectonic movement and opened a passage between North and South America. At the time, some 3.5 million years ago, there was a concomitant expansion of many North American mammal taxa into South America. There was also some expansion in the opposite direction (the armadillo and opossum are familiar examples of the latter), but the arrow of movement ran primarily from north to south, as many more North American species expanded into South America than the other way around. There was also an increase in extinction rates among South American taxa after North and South America became joined.

Paleontologists have long recognized that the Great American Interchange is a good example of a phenomenon that is equivalent to geodispersal, though it has not been described explicitly in that light. Traditionally, the forces believed to have motivated this event were related to competition because of what appeared to be the relative success of the North American mammals at the expense of those of South America (e.g., Darwin, 1859; Wallace, 1876; Matthew, 1915, 1939). However, recent investigations have suggested a less significant role for competition as the driving force for that event. For a more detailed discussion of the Great American Interchange the interested reader should see Webb (1978) and Vrba (1992). Several paleontologists including McKenna (1975), Hallam (1977, 1983, 1994), Fortey (1984), and Jablonski *et al.* (1985) have described numerous examples of biogeographic patterns in fossil taxa that are equivalent to what I am referring to here as geodispersal.

Vicariance biogeographers, who typically work with the extant biota, also have discussed biogeographic phenomena akin to geodispersal, and I shall mention some of them shortly. However, in spite of this general acknowledgment, they have rejected the relevance of any type of range expansion to historical biogeography. They have for the most part been vehemently opposed to even considering the possibility that range expansion as a process shapes biogeographic patterns, because they treat all types of dispersal as equivalent. Their opposition can be attributed to two factors—one valid and one not. The first has already been discussed, and it has been recognized that traditional dispersal is not governed by large-scale geological or climatic events, nor does it produce congruent biogeographic patterns among different clades. The second factor arose as an assumption that all types of range expansion are equivalent, and that none of them can be associated with congruent events seen in other clades.

In the case of geodispersal this is simply not true, as I have demonstrated above. The rejection of all types of dispersal as subjects of interest to vicariance biogeography is the reason that they have failed to develop analytical techniques that uncover episodes of geodispersal. This is problematic, for any analytical approach that fails to take geodispersal into account will have serious shortcomings. As geodispersal is such a pervasive phenomenon, this attitude has limited the ability of vicariance biogeographers to retrieve well-resolved and meaningful biogeographic results in their studies, and I believe has, in turn, seriously limited practitioners in historical biogeography.

The paradox in the fact that vicariance biogeographers first recognized the phenomenon of geodispersal and then roundly rejected its relevance warrants a few examples that demonstrate this recognition. One vicariance biogeographer who deserves mention for identifying a process akin to geodispersal was

Rosen, who noted that, "there are two processes that have molded biotas into their present configurations: large-scale dispersal to produce widespread ancestral biotas and later allopatric speciation events which have fragmented the ancestral biotas into their present highly subdivided states" (Rosen, 1978, p. 160).

Platnick and Nelson also (1978) identified two types of dispersal—traditional dispersal and another type that can lead to several species having broad distributions. The latter involves range expansion when a barrier is removed and is of course equivalent to geodispersal. However, they treated it as essentially another indication of vicariance, because for clades to differentiate by vicariance they have to have been broadly distributed in the first place. They subsequently reiterated this (Nelson and Platnick, 1981) noting that,

> dispersal is vicariance in disguise...the reason is that the postulated dispersal takes place prior to the appearance of the barrier and prior to the fragmentation of the range of the ancestral species. The effect of the postulated dispersal is only the creation of primitive cosmopolitanism (a requirement of the vicariance model). (In Brundin, 1988, p. 356)

They (Platnick and Nelson, 1978; Nelson and Platnick, 1981) are only correct in equating geodispersal with vicariance if clades in a biota attain a broad distribution just once, during their original occupation of a region, and then subsequently undergo vicariance. Their portrayal will be wildly inaccurate if clades undergo geodispersal and subsequent vicariance numerous times during their history. Throughout Earth history geographic barriers have formed only to subsequently fall, so there is no reason to believe that most biotas have been shaped by a single episode of vicariance that was preceded by one period when all ancestral species were broadly distributed. Instead, geographic ranges of species within clades should oscillate between broad and narrow distributions.

Platnick and Nelson (1978, p. 7) also invoked a geological example that produced what I have defined as geodispersal—the collision between India and Australia during the Cenozoic era—and they termed it "biotic- dispersal." I consider this term inaccurate and potentially confusing because the range expansion is not caused by biological factors at all, except in a trivial sense; rather it is related to geological change. Therefore, I prefer the term geodispersal. However, what is important is that the existence of the same phenomenon, congruent episodes of range expansion reflected in several taxa, was recognized by Platnick and Nelson (1978), two of the primary architects of the vicariance biogeography approach, although they were generally virulently opposed to invoking traditional dispersal in any explanation whatsoever because they did not think that it could produce congruent biogeographic patterns. They also believed that different groups of organisms could not show similar patterns of dispersal, which, owing to the occurrence of geodispersal, is not true.

Several other authors recognized the existence of a phenomenon akin to geodispersal. Cracraft (1988, p. 233) and Noonan (1988, p. 377) noted that geographic barriers preventing the movement of organisms cyclically appear and disappear, such that there could be vicariance and also, subsequently, (geo)dispersal. Wiley (1988a, p. 271) also noted that there could be dispersal involving entire biotas, which is, in effect, geodispersal. Brundin (1988, pp. 348, 362–363) classified dispersal into jump dispersal over barriers (traditional dispersal herein), which he believed was rare, and also what he referred to as range expansion owing to the disappearance of barriers. He believed that "vicariance and range expansion have constantly alternated, bringing about intermittent interchange between biotas of different parts of the world" (Brundin, 1988, p. 356–357).

Bremer (1992), Ronquist (1994, 1998), and Hovenkamp (1997) were also strong advocates of the idea that oscillations between vicariance and range expansion have powerfully influenced biogeographic patterns throughout the history of life. Each of these perspectives differs significantly from what Rosen and Platnick and Nelson had discussed (Rosen, 1978; Platnick and Nelson, 1978; Nelson and Platnick, 1981). This is because the latter three researchers believed that there was only a single episode of geodispersal early in the history of a biota, whereas the former seven scientists were prepared to invoke several episodes of geodispersal.

All of these extensive writings suggest that historical biogeographers would be receptive to the idea that episodes of range expansion can occur concurrently in several different clades. Range expansion can also be congruent, just like vicariance! Thus, it can find explanation as a general phenomenon influencing historical biogeographic patterns. Although the earliest ideas on what I have termed geodispersal were developed by geologists with a paleontological background, many phylogenetic systematists who have emphasized the strength of vicariance explanations of biogeographic patterns also recognized that a phenomenon akin to it can occur.

Perhaps the unwillingness of vicariance biogeographers to be swayed to a belief in the general validity of geodispersal had to do with the perceived lack of an appropriate method to study it. However, as I shall discuss more fully below, such methods do exist and they are very similar to some of the commonly utilized phylogenetic biogeographic techniques. I would argue that the key element of historical biogeographic explanations is not an emphasis solely on vicariance, but rather an emphasis on congruence, which should take the form of independent clades showing similar changes in their geographic distribution associated with cladogenetic events. These changes could involve range contraction as groups evolve, but they could also involve range expansion.

6.6. Integrating Vicariance and Geodispersal with Plate Tectonics

This book does not deal with issues relating to the discovery of plate tectonics because these have been adequately considered elsewhere from an historical perspective by Hallam (1981), from a geological perspective by Uyeda (1978), and from a biogeographic perspective by Brown and Lomolino (1998), and there is no need to repeat these extensive treatments. However, the discovery of plate tectonics was an extremely important event in the history of the Earth sciences, and the earliest ideas related to plate tectonics, those bearing on the theory of continental drift, were based on biogeographic patterns involving the distribution of fossil organisms on now disparate continents. The crucial role biogeography played in the development of plate tectonics makes it worthwhile to integrate the biogeographic patterns I've been discussing into a plate tectonic framework.

The existence of plate tectonics implies that the Earth is broken up into a series of large plates, some of which include continents. The geometry of the Earth's plates has been incessantly shifting over the face of the globe, with different plates and land masses separating, sliding past one another, or coming into contact at plate boundaries. In a simplistic sense plates can be thought of as rigid bodies sliding over an underlying fluid layer termed the asthenosphere. Average plate motion is on the order of a few centimeters a year, which is not much, but when extrapolated over millions of years it can result in very large magnitudes of displacement.

Plate boundaries can be situated along the edges of continents, within continents, or within oceans, and there is a dramatic variation in plate size. Further, the Earth's plates, in addition to being in motion, are also in a continual state of flux in terms of what is happening at their margins. New oceanic plate material is continually being generated, and old material is being destroyed and brought back into the Earth. New oceanic seafloor is generated at ridges, places where hot magma wells up from within the Earth to the surface. These ridges or zones are elevated relative to the seafloor, and in places may even be above the ocean surface, as in the case of Iceland. The regions in the oceans where oceanic plates are brought down into the Earth or get subducted tend to be very deep narrow bands, referred to as trenches. Basically the continents can be thought of as floating high on the asthenosphere.

Generally speaking, continental material does not get subducted. Continents do collide, as, e.g., the collision between India and Asia in the Cenozoic era, which has driven the uplift of the Himalayas. Continents can also rift apart. Eastern Africa is a region where continental crust is separating. Further, different pieces of continents can slide past one another (in some sense, a type of collision). The most famous example of this phenomenon involves the interaction between the Pacific plate and the North American plate in southern California. There, a tiny sliver of the United States, lying to the west of the San Andreas fault, is moving northwestward relative to the rest of North America.

For our purposes, i.e., understanding the distribution of organisms and how they have evolved as plate tectonics has caused their geographic distributions to change, we need to be concerned with only two different kinds of tectonic events. The first is tectonic rifting, which can separate joined regions, isolating their respective terrestrial and marine biotas (see Fig. 3). We would expect that tectonic events of this type would lead to vicariance in terrestrial biotas and in marine biotas that are distributed around the margins of continents and live close to shore. Groups that were affected by such events should show a classic pattern of vicariance, diversification associated with a contracting range. In other words, as a continent rifts apart a clade will diversify, with each diversification event occurring in a more narrowly restricted geographic range than the original ancestral range of the group. Geographic ranges will tend to become bisected in the sense that as one proceeds up the tree the geographic range occupied by the putative ancestral species tends to shrink (see Fig. 6).

The second kind of plate tectonic event that concerns biogeographers is tectonic collision, which, by bringing tectonic plates together, can unite isolated regions and thus potentially both marine and terrestrial faunas that had been separate (see Fig. 4). When this happens one would expect to see a sudden expansion in the ranges of species in different clades, i.e., geodispersal. These collisional events can be profound in the long term, as in the case of the collision between India and Asia, or they can be more subtle, as when different regions touch as the plates they are on slide past one another.

The one other type of event that can greatly influence biogeographic patterns in both terrestrial and marine faunas is sea-level change, which is sometimes caused directly by climatic changes. When global temperature drops, more seawater is stored in the polar caps and global sea-level tends to drop, and when global temperature rises, the polar caps melt and sea level rises. However, the global climate system, the size of the polar ice caps, and thus sea level are all related to plate tectonic events. For example, the closure of the Isthmus of Panama, discussed earlier, had a major effect on oceanic circulation, which caused global temperature to fall and the polar ice caps to grow. Also, the increasing isolation of Antarctica in the Cenozoic, driven by rifting that moved South America and Australia away from the South Pole, also led to a drop in mean global temperature and the formation of the Antarctic ice sheet (Stanley, 1998).

Sea-level rise and fall can also be more immediately related to plate tectonics as with geological processes that mediate the sizes of the midocean ridges, which extend around the globe and are the sites where new plate material is generated. The great geographic extent of these ridges and their significant height relative to much of the seafloor mean that volumetrically they are quite large. This immense system, being mostly under water, can be thought of as displacing that seawater. It turns out that the volume of the ridges, and thus the volume of seawater that they displace, varies. During certain periods new plate material is generated more rapidly than at other times, so the volume of the ridges is greater, displacing more seawater, which would cause the global sea level to rise. By contrast, when new plate material is generated less rapidly, the volume of the ridges shrink, displacing less seawater, so that the global sea level falls. The absolute magnitude of the change in sea level that is due to these variations in ridge size is in fact greater than the changes that the expansion and contraction of the polar caps can cause. However, the polar caps can change size and influence sea level more rapidly than the ridges, so climatic changes cause more rapid rises and falls in sea level.

It is easy to develop scenarios that show how sea-level change can act to influence biogeographic patterns and concrete examples will be provided in later chapters. However, we can imagine a decline in global sea level, due either to climate change or tectonic processes, that can cause formerly inundated land masses to become emergent and elevated above the ocean surface. For example, if sea level were to fall sufficiently the Bering Strait separating Asia and North America would become emergent, allowing terrestrial Asian species to expand their ranges into North America and vice versa (McKenna, 1983). At the same time, sea level fall often has the opposite effect on marine biotas.

In the same scenario populations of marine organisms that were formerly continuous would become isolated on either side of the strait. They would then differentiate and eventually undergo speciation via vicariance. Because the sea level changes that are climatically mediated occur more rapidly than those that are tectonically mediated, we might think that they would play more of a role in influencing patterns of range expansion and vicariance and speciation than changes in ocean ridge volume. However, this would not prove true if the climatic changes occurred so rapidly that they caused geographic barriers to rise and fall too quickly to allow sufficient time for isolation, speciation, and vicariance to occur.

6.7. Relating Earth History and Evolution: General Perspective

It is clear that the processes of vicariance and geodispersal potentially tie evolutionary change to geologically mediated events, be they plate tectonics or climatic change. The pervasiveness of the phenomenon of allopatric specia-tion, discussed in Chapter 5 and in Mayr (1942, 1982), Eldredge (1971), Eldredge and Gould (1972), Lynch (1989), and Brooks and McLennan (1991), is one reason for us believe that the tie between evolution and Earth history events such as geological and climatic change is strong. There are also larger-scale patterns that illustrate the strength of this connection. First, there is the excellent overall correlation between the geometry of the Earth's plates and biological diversity (Valentine and Moores, 1970, 1972; Jablonski *et al.*, 1985). When the continents are fragmented and dispersed biological diversity is generally high because regions tend to be separated by geographic barriers, which encourage speciation and delineate species ranges. The end result will be more areas of endemism. By contrast, when the continents are joined together, diversity tends to be relatively lower. When cratons assume such a geometry, there should be fewer potential geographic barriers, fewer opportunities for speciation, and fewer areas of endemism.

Another reason to believe that the tie in between Earth history and evolution is strong is the perception that key episodes in the history of life correspond to major events in Earth history. For example, the initial diversification of animal life, during the Cambrian radiation, when 20 phyla appeared in a geologically short period of time, was one of the most profound events in the history of life. Because this interval also corresponds to a time of major geological change, many authors have suggested that Earth history events played an important role in motivating the Cambrian radiation (e.g., Knoll, 1991; Signor and Lipps, 1992; Dalziel, 1997; Lieberman, 1997). It turns out that this interval corresponds to a time of major events in Earth history that involved changes in the relative positions of the cratons. These were all joined in the Neoproterozoic but then began splitting apart, which created numerous opportunities for vicariance and speciation. There were also major climatic changes at this time. In addition to the dramatic diversification of major animal groups at high taxonomic levels during the Cambrian radiation, speciation rates were also relatively high (Lieberman, 1999a; Lieberman, 2000), which might be explained by invoking the tectonic events that occurred at the time.

Although the Cambrian radiation stands alone as a fundamental event in the animal lineage, there were many other times throughout the Phanerozoic in which Earth history change and biological change appear to be conjoined. The Ordovician radiation, another time of major biological diversification corresponded to a time of major tectonic change (Miller and Mao, 1995). The Eocene–Oligocene transition, a time of profound climatic change, was a fundamental event for mammal lineages (Prothero, 1994). McKenna (1983) described other cases throughout the Cenozoic where geological events joined and later separated the European and North American continents, leading first to an exchange of mammals between these regions and subsequently to diversification and endemicity. Finally, on smaller spatial scales, Bayer and McGhee (1988) and Geary (1990) described how the evolutionary changes in the gastropod fauna of the Black Sea were driven by the tectonic motion of Africa relative to Eurasia, which isolated the Black Sea from other marine ecosystems and led to dramatic evolutionary changes in the snails of the region.

Many other instances of coupled tectonic and climatic change forcing evolutionary change are known. Perhaps one of the most profound, from our own perspective, was the cooling in eastern Africa during the Neogene. Driven partly by immediate tectonic factors such as the closure of the Isthmus of Panama and the alteration of global oceanic circulation patterns, this cooling appears to have precipitated factors that led to the origin of our own lineage, the hominid genus *Homo* (Vrba, 1992, 1996). As I noted earlier, this event also led to the Great American Interchange. The tectonic and ensuing climatic changes also influenced the molluscan faunas of the eastern Pacific and western Atlantic. After the isthmus closed there were differential patterns of extinction, with species in the western Atlantic suffering higher degrees of extinction than those in the eastern Pacific [see Vermeij (1978)]. Although this climatic cooling was immediately precipitated by the closure of the isthmus, it was amplified by other even earlier tectonic events that had already led to a drop in global temperature, e.g., the Antarctic continent had become isolated over the pole, leading to the generation of a circumpolar Antarctic current (Stanley, 1998). This sequence of events illustrates the close coupling between geological and climatic change and biological change.

Finally, Wells *et al.* (1999) recently demonstrated that tectonic activity and increased generation of new plate material at the midocean ridges can act to enhance the biological productivity of the oceans. This might be another reason that times of major tectonic change, specifically seafloor spreading and rifting, correspond to key episodes in the history of life. From this perspective, the Cambrian radiation becomes an even better example of a biological event governed by Earth history phenomena. Not only would the increased rifting during the Cambrian radiation have contributed to elevated speciation rates, but it might have enhanced biological productivity, thus providing further opportunities for evolutionary diversification.

6.8. Limits of Resolution in the Fossil Record, Our Ability to Identify Paleobiogeographic Patterns, and Conclusions

Thus far I have discussed a relatively straightforward framework for relating plate tectonic events to biogeographic events. Historical biogeography can be viewed as the study of two related subjects: the use of biogeographic patterns to determine (1) the role that Earth history plays in evolution; and (2) to track sequences of geological events. These two approaches are described below, but we have to recognize that limits of resolution in the fossil record affect our ability to pursue them. For example, researchers might recognize that there was a time of major tectonic and/or climatic change in the geological record and then want to look at biogeographic patterns to see if the extensive Earth history events seem to have had a prominent influence on patterns of evolution. If such influence could be demonstrated, it would show that Earth history events can have a powerful effect on the evolution of life. (How this can be done will be discussed in Chapter 9.) Biogeographic analysis emerges as one of the best ways to test this though there are others; e.g., establishing that large-scale volcanism was correlated with a large-scale biotic crisis would demonstrate this cause-and-effect relationship [see, e.g., Renne *et al.* (1995)].

Geological researchers also might want to piece together the timing of major tectonic and climatic events, and biogeography would be an important part of such a research program, though there are other techniques as well, including the search for the presence of certain lavas and the recovery of radiometric dates, which indicate when they formed. Depending on the rock type, these might be associated with continental rifting or certain kinds of continental collision. There are also characteristic changes in rock types seen through time, which indicate different kinds of continental collision or rifting (Bond *et al.*, 1984; Pelechaty, 1996). Other available techniques include tracking the paleomagnetic stripes on the ocean floor, which allows the relative sequence and geometry of seafloor spreading events to be worked out moderately well. Simply by subtracting these stripes away one can determine the position of a plate at a certain time and thus reconstruct the movement of plates and tectonic events. Paleomagnetic information can also be used to constrain the original latitude where a continental rock formed, which can figure prominently in reconstructions of the movement of the plate that the rock comes from, which helps to identify tectonic events.

However, each of these paleomagnetic techniques has shortcomings. For instance, techniques that reconstruct the paleolatitudes where continental rocks formed are sadly mute as to the paleolongitude where they formed. Further, we cannot use magnetic stripes on the seafloor to trace tectonic events prior to 150 million years ago because there is little if any intact seafloor older than this in existence. Thus, in many cases, information from biogeography becomes extremely important.

In Chapter 3 I discussed the limits of temporal resolution in the geological record. These limits influence our ability to conduct paleobiogeographic research. Let us consider the first type of paleobiogeographic research I described, which focuses on how geological changes impacted biological entities. If the geological record is incomplete, many tectonic events that geologists might have picked up by geological means, cannot be discerned, which would obviously limit our understanding of the record of tectonic events. There may be no geological evidence for some tectonic events, so their influence cannot be predicted, yet they will have influenced biogeographic patterns. These "hidden" events would obfuscate matters, because predictions would be based on inadequate sampling. We must also realize that even when techniques such as radiometric dating or the analysis of changes in rock types can be used, there is always going to be some imprecision about the timing and number of tectonic events that can never be resolved. These uncertainties further limit our ability to tie together Earth history events and evolution.

The incompleteness of the geological record also engenders problems for the study of the sequence of tectonic and climatic events. Many of the species that might have been around during the time interval of interest may not have been preserved in the fossil record, which could affect our understanding of evolutionary events and biogeographic patterns. Moreover, changes or oscillations in the geographic ranges of organisms in response to tectonic events may have been too rapid to have been preserved in the fossil record, and would thus not be visible. It is also conceivable that the tectonic and climatic changes that can mediate geodispersal and vicariance recur repeatedly, and closely on the heels of one another. This, in combination with an incompletely preserved complement of species from the time interval of interest, would result in biogeographic patterns that are either not easily interpreted or easily misinterpreted.

Another serious problem involving limits of resolution in the fossil record relates to philosophical views about the nature of biogeography and dispersal. For instance, we can imagine a case where a geographic barrier to the free movement of organisms has fallen. The expected result would be range expansion or dispersion in several different lineages with species migrating into new regions. Only later, if a new barrier emerges within the expanded ranges of these species, would we expect the populations in both the original and the new area to differentiate via vicariance.

Let us now imagine that in several clades such a vicariance-style event follows a dispersal event such that the phenomenon of geodispersal can be invoked. Now, we suppose that the fossil record is somewhat coarse, such that our limits of resolution do not allow us to see the dispersal events immediately after the fall of the geographic barrier. However, we further suppose that the fossil record is not so coarse that we do not see the final by-products of dispersal followed by vicariance, which would be new species appearing in a new area (see Fig. 9). If the observed phenomenon, witnessed in

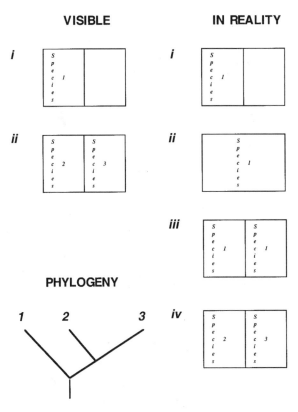

FIGURE 9. Limits of paleontological resolution and interpreting paleobiogeographic patterns in the fossil record. First, consider the "Visible" column: In the fossil record we imagine a single species confined to a single area. Later, we see two new species, 2 and 3, separated by a geographic barrier. Species 2 is found in the same area as species 1. Species 2 and 3 share a sister-group relationship and these two in turn share a sister-group relationship with species 1. This situation matches the definition of traditional dispersal. However, if this pattern is seen simultaneously in several clades, it more likely matches what is seen in the "In Reality" column, which is geodispersal. Here, a single species is confined to an area. Then, a barrier falls and the species expands its range. Later, the same barrier rises again. Eventually, the isolated population differentiates via vicariance and undergoes speciation.

the fossil record, a new species in a new area (actually several new species each from a different clade in that new area), were interpreted literally, for each clade it would fit the definition of traditional dispersal or jump dispersal described above, i.e., expansion into a new area (over a geographic barrier) with concomitant diversification.

However, traditional dispersal would not really describe what had happened. The philosophical approach that will be taken here is that traditional dispersal does not occur congruently in several different clades. It involves the chance passage of a single lineage over a barrier, and then that lineage subsequently diversifies. If we find in the fossil record that new species appear in a new area in several different clades, it is extremely improbable that this could have been produced by traditional dispersal. However, it could easily have been produced by dispersion followed by the emergence of a barrier with subsequent vicariance—what I have termed geodispersal. I propose that the criterion to be used to assess whether traditional dispersal or geodispersal has occurred should be congruence. If several clades show the same pattern of range expansion, the only way we can account for it is through geodispersal. However, in the fossil record, which is not complete, the intervening stages that lead to geodispersal may not all be visible.

Chapter 7

Defining Areas in Paleobiogeography

7.1. Introduction

The discipline of biogeography, as has been emphasized repeatedly, was developed because the same species are not found everywhere. Different regions do not have the same complement of species and, moreover, the climate alone is not sufficient explanation for why species are found where they are. As part of the discussion of biogeography and paleobiogeography, several other topics were introduced and discussed, including the role that vicariance and allopatric speciation play in influencing biogeographic patterns: they produce closely related species separated by geographic barriers. The role of geodispersal as a mechanism that alters the distribution of species by moving them from one region to another was also discussed, and it has been argued herein that this is an important biogeographic process that will lead to patterns of different species occurring in different regions or areas of the globe. Thus far, however, in this book the terms regions and areas have been treated in a fairly nebulous fashion, connoting something to do with geography and position on the Earth's continents or in the Earth's oceans. However, as biogeography is the study of how life is distributed on Earth, a more rigorous definition of regions and areas is required.

What is the bast way to view regions and areas in a biogeographic context? The fact that different species are found in different areas was first recognized by Buffon in the 18th century, and this was later generalized by de Candolle père into a statement that groups of species are often restricted to individual regions (Nelson, 1978). In this context, de Candolle père referred to endemic genera, which are genera in which all of the constituent species are confined to a single region. These so-called endemic forms are not uniformly distributed across the Earth, but are clustered in different places (Brown and

Lomolino, 1998). The regions housing them differ in size, and some have more species than others. This latter fact may be partly related to the phenomenon of latitudinal diversity gradients that I discussed earlier when dealing with the sub-discipline of ecological biogeography, but we have not found a one-to-one correspondence between ecological and historical biogeographic patterns. The regions that house endemic species are called areas of endemism, and at least since the time of Latreille (1822) and Lyell (1832) it has been known that they are generally surrounded by geological or climatic barriers.

In considering which regions we should select to answer questions of biogeographic significance, we might envision many concepts as to what constitutes a region. On the one hand, the tropics of the Americas, Asia, and Africa, because of their great diversity of species, might be an interesting region to study in regard to large-scale ecological biogeographic questions. On the other hand, an ecologist might burn 10 m^2 patches, eliminating all living things, and then look at how each of these patches is subsequently colonized. Each patch could be defined as a significant region that could be studied within an ecological biogeographic framework. We could even draw random star shapes across a map and identify each of them as a significant region, though we might expect that unless there was something about the evolutionary or ecological process that produced star-shaped distributions, such regions would be uninteresting from a biogeographic perspective.

Within an historical biogeographic framework we also need to consider what defines an area of endemism. It is likely that areas so defined will differ from the regions mentioned above because historical biogeography focuses on how patterns of speciation within and across clades are related to geological and climatic processes. The fact that species tend to be concentrated in distinct regions separated by geographic or climatic barriers indicates that some geological or climatic process impacted on the evolution of the biota in any particular region, which shared a common history to the exclusion of biotas outside of the region. How should areas of endemism be defined? Specifically, how should we view these entities and why do they exist? What is their underlying ontological basis? How do we recognize them? In other words, there is also an epistemological question here. In many ways, the issue of defining areas involves some understanding of the ontology and epistemology of areas, and thus defining them has many analogies to the problem of defining and identifying species in biology [see Eldredge (1985*a,b*)]. The analogy between species and areas of endemism, which shall be explored here, is part of the broader analogy between biogeography and phylogenetic systematics that will be explored more thoroughly in the next chapter. Here, I first discuss various species concepts from the fields of systematics and evolutionary biology and then relate them to the definition of areas in biogeography and paleobiogeography. Although Hovenkamp (1997) has recently questioned the existence of areas of endemism (this rejection parallels the refusal of some authors to accept the existence of species, which is discussed more fully below), there seems to be overwhelming support for it from many ancillary lines of evidence, even in a preevolutionary framework (see Chapter 4). This brings to mind the way Mayr (1942, 1963, 1982) demonstrated that species exist, by marshaling extensive evidence, including the interesting fact that natives of New Guinea with no training in systematics identify the same basic avian species there that ornithologists do. The confirmation of the pervasive phenomenon of geographical speciation, which associates evolutionary divergence with disjunctions caused by geographic barriers, would also lead one to suspect the existence of areas of endemism.

7.2. Species Concepts: Ontology and Epistemology

Tying the definition of areas to that of species may seem unwise or even cause for dismay, because one of the greatest debates in the history of biology centers around what, if anything, is a species. Moreover, based on current opinions about the nature of species, it would be safe to say that it has yet to be resolved. Still, the area problem may prove less troubling: As the question of what is a species has been so thoroughly discussed, we can choose the aspects of that debate that have, over the ages, become more clear-cut, and apply them to the delineation of biogeographic areas of endemism. Part of the confusion about the nature of species has to do with the difference between ontology, i.e., what are they, and epistemology, i.e., how do we recognize them (Eldredge, 1985a,b 1989a,b).

Many authors have discussed important and exciting approaches to the species problem, but in the interests of brevity and clarity, I will concentrate on those whose work shed light on the definition of areas. The most relevant concepts are those of biological species [and related specific mate-recognition system (SMRS)], evolutionary species (and the related phylogenetic species), and cladistic species. A brief background will be given for each of them, and then the interesting conceptual approach of de Queiroz and Donoghue (1988, 1990), which provides a means of adequately resolving the nature-of-species debate, as well as potentially the nature of areas, will be discussed.

The first of these ontologies, the biological species concept, is the proposition that species are actually or potentially interbreeding groups of organisms (Dobzhansky, 1937; Mayr, 1942). Viewing a species as a collection of interbreeding or potentially interbreeding organisms implies that populations of a species are part of a nexus that shares genetic material. In the long term, such species might be expected to often, but not always, share some common evolutionary history. The "not always" qualifier is necessary because potentially interbreeding populations may be separated for long periods of time such that they have separate evolutionary histories, but may still be able to interbreed, when they are subsequently brought back together again.

Another similar ontology and definition of the nature of species is the related perspective that species are defined by a SMRS (Paterson, 1985) such that members of the species recognize one another for the purposes of interbreeding and do not recognize other organisms. In this perspective, there are behavioral, morphological, and/or physiological characters that enable successful breeding between animals of the same species. If we could discover some or all of these characters, as Vrba (1984b) has done for some of the living and fossil antelopes of Africa, then we would understand the nature of the SMRS in a particular group, and how it evolved through time. Characters can also help scientists recognize species: if one uncovers a distinctive SMRS, one has found a species. This means they can also be used in an epistemological definition of species. However, at base, the SMRS is a statement about the nature or ontology of species—that they are a reproductively associated collection of organisms and sit as self-defining wholes, independent of other whole species (Lieberman, 1992).

The biological species concept, which has been accepted by many evolutionary biologists since the formulation of what has been termed the Darwinian synthesis in the 1930s and 1940s [see Eldredge (1985)], and the related SMRS concept have been criticized by systematists, who have argued that it should be abandoned (e.g., Sokal and Crovello, 1970; Donoghue, 1985; Nelson, 1989). Their criticism is based in part on the fact that the biological species concept is hard to operationalize. We cannot always know, especially with extinct or geographically isolated organisms, which populations could or did interbreed, and this would be an epistemological argument for the concept's inadequacy.

Further, it has been convincingly demonstrated that biological species may not correspond to monophyletic entities that share a history of common descent from a single common ancestor (Rosen, 1979; Donoghue, 1985; de Queiroz and Donoghue, 1988; Nelson, 1989). That is, in some biologically defined species that consist of several populations, some of the populations may be more closely related evolutionarily to populations in other species than to populations within their own, such that they would be paraphyletic. This comes about because populations can be isolated from one another for long periods of time, but subsequently reunite and reproduce. Sometimes populations that are distantly related evolutionarily can still interbreed while others that are more closely related in that respect cannot. Some researchers have argued that the failure of biological species to correspond to mono-phyletic entities is cause for the biological species concept to be abandoned.

This second argument against the biological species concept and the SMRS is more about the ontology of species. Some proponents of this argument suggest that the way to view species (ontologically) is as collections of monophyletic entities that would share a unique evolutionary history and a pattern of descent from a common ancestor. Such an ontology is found in the related evolutionary species concept of Simpson (1961) and Wiley (1978, 1981), according to which species are held to be phylogenetic lineages with their own distinct evolutionary tendencies, isolated from other such lineages. Although Simpson (1961) did not use the term monophyletic to describe such lineages, their monophyly would be a necessary by-product of such a definition. A potential problem with this ontology of species is that monophyletic lineages exist at several hierarchical levels, and no one particular level can be dignified as connoting special significance in terms of shared common history. Further, in some sense, shared patterns of common descent and the divergence of lineages must be related to cessation of reproduction, so perhaps an interbreeding criterion is important to the definition of species.

If we ignore these problems for the moment, this concept would be a potentially valid ontological view of species, and with this perspective, species could be recognized or identified by their shared possession of unique character states. In other words, epistemologically, species could be viewed as possessing synapomorphies. In this respect, this species concept is similar to that of phylogenetic species defined in Eldredge and Cracraft (1980), as well as by later authors, in which a species is described as the smallest collection (minimally a male and a female in sexual species) of organisms that can interbreed, and is defined by one or more uniquely shared cladistic characters. In this perspective, systematic characters can be used to identify species, although they have some real existence independent of these characters.

In fundamental opposition to each of these views is the cladistic species, of Nelson (1989), who tried to define species as equivalent to other taxa. To him, species, like other taxa, are homologies (would connote any type of character that is shared) and have homologies (e.g., Nelson, 1989, p. 279). His approach is unique because it is strictly an epistemological definition of species, which he says are only their characters, and he has attempted to define characters independent of a notion of common descent. Although he was unsuccessful in this regard because implicit in his definition of homology was some statement about reproduction and inheritance (Nelson, 1989, pp. 281, 282), his definition still serves as an interesting counterpart to the species definitions discussed thus far: it was designed to be a definition of species that was independent of any ontology. This is part of Nelson's (1989) general pattern cladistic framework. To him, all that defines species or other taxa is their ability to be recognized, i.e., their epistemology, and they have no existence independent of their synapomorphies.

De Queiroz and Donoghue (1988, 1990) provided several interesting perspectives on the nature of species. Perhaps the most important, in terms of the discussion here, is their contention that species can have a real existence in nature through their status as communities of interbreeding organisms and their shared history of common descent. That is, ontologically, species can be defined either as reproductive communities or as monophyletic entities. Each of these definitions has its relative strengths, and both are important ways of viewing the evolutionary process.

On the other hand, de Queiroz and Donoghue (1988, 1990) rejected Nelson's (1989) view that taxa exist only because of their shared characters, arguing that higher taxa have some ontological status that derives from their shared pattern of common ancestry, i.e., evolution by common descent. Characters may be used to determine this pattern, but the taxa themselves and the shared pattern of common ancestry exist independently of the characters. In support of this, they cited Hennig (1966, p. 8), who said characters "are not themselves ingredients of the definition of the higher categories, but aids used to apprehend the genetic criteria that lie behind them." [This idea actually goes back to Linnaeus' *Philosophia Botanica*, where he wrote that it is the genus that gives the characters. That is, that there is some real entity out there, and we can use characters to try to discover it, except that in this context the real entity is the species rather than the genus (Mayr, 1982).]

At the species level, de Queiroz and Donoghue (1988, 1990) take a broader ontological view, arguing that species may be viewed as systems sharing a pattern of common ancestry, like higher taxa, or they may be viewed as communities of reproductively compatible organisms. Depending on the researcher's approach, either of these views may have heuristic strengths. Eldredge (1989*a,b*) echoed a similar sentiment. It is clear that the views of de Queiroz and Donoghue (1988, 1990) and Eldredge (1989*a,b*), in contradistinction to Nelson (1989), emphasize the reality of species, and this reality, which may be governed by several different processes, exists independently of the synapomorphies that might be used *a posteriori* to identify them.

The viewpoint of Nelson (1989) that species have solely epistemological meaning seems intellectually narrow, and it is not the one I advocate here. This epistemological view allows little prospect for integrating systematics with evolutionary biology and, moreover, it may have logical flaws, as I intimated above. Without the connection between systematics and evolutionary biology by means of common descent, systematics becomes a dry and meaningless pursuit. Some link needs to be maintained in order to demonstrate the critical relevance of systematics as a general discipline and as a discipline that can make important contributions to the field of evolutionary biology.

7.3. The Individuality of Species

Part of the debate about the reality of species has also focused on whether or not species are classes or individuals. The notion of the individuality of several entities within the hierarchy of life was introduced in Chapter 2. There I noted that biological individuals in the sense of Ghiselin (1974) and Hull (1976, 1978, 1980) have a history, with a birth and a death point; they are spatiotemporally localized, and they have some stability or constancy during their existence. Species fit these criteria. Their birth point is speciation, and their death point is extinction. In addition, species are held together or given cohesion by reproduction among their component populations and organisms. Further, they have the quality of more-making or speciating, where more entities of like kind are created.

Individuals in this sense are to be distinguished from classes [see, e.g., extensive discussion in Ghiselin (1974), Hull (1976, 1978, 1980), Wiley (1981), Vrba and Eldredge (1984), (Eldredge, 1985*a,b*, 1989*a,b*), de Queiroz and Donoghue (1988)], which are defined by listing the attributes that characterize them. Nelson's (1989) contention that species are like other taxa, and simply are the characters they have, the purely epistemological approach to species, really corresponds to the view of species as classes. By contrast, the way de Queiroz and Donoghue (1988, 1990) have defined species has more in common with the species-as-individuals perspective. To them, species can have a history and a shared pattern of common descent, and they can be held together by reproductive cohesion. The species-as-individuals debate is really about whether or not species have ontological status, and how that status relates to evolutionary processes.

7.4. Translating the Debate about Species to the Debate about the Nature or Ontology of Areas

How can we extend these ideas on the nature of species to the nature of areas? We might first think about the different species concepts discussed above, and then return to the notion of whether or not areas might be individuals or classes. Implicit in the discussion of areas and historical biogeography was the notion that important areas to consider were areas of endemism, i.e., areas characterized by unique taxa, which can be populations, species, or higher taxa. However, owing to the limits of resolution in the fossil record, when considering paleobiogeographic patterns it is probably necessary to restrict our focus to species or higher taxa, because population-level biogeographic differentiation is likely to be too ephemeral or too local to be readily visible there.

Two major perspectives on species have been discussed above. First, the preferred ontological view, in which species have a real existence in nature. The two reasons offered for this ontological existence were reproductive community and common descent, and de Queiroz and Donoghue (1988, 1990) provided evidence that both were legitimate. The second major perspective was that of Nelson (1989), which is the strictly epistemological approach, i.e., species are merely the characters that define them, which was rejected here as invalid.

Either of these perspectives could be extended to the definition of areas. We first imagine the consequences if we take Nelson's (1989) approach. In this case, areas would simply be epistemological phenomena, their existence derived from their possession of one or more unique species, which would define them just as Nelson (1989) felt that unique characters define and give existence to a species. Then, areas would have no reality other than the fact that they housed some unique set of taxa. A similar type of epistemological approach would define areas solely by a set of geographic barriers. Either of these views would make it impossible to consider areas in the context of certain processes that might have governed their existence and how they changed through time.

These processes, as was suggested above, are likely to be either geological or climatic. If the immediate connection between geological processes and areas is sundered there is no reason to even pursue biogeographic research, which, as I noted earlier, should be dedicated to looking at the coevolution of the Earth and its biota. A strictly epistemological definition of areas would imply that this kind of study has no value as the regions that have unique complements of species (or which are delineated by geographic barriers) have no real existence other than that complement of species. They would be merely classes defined by their shared possession of certain attributes.

The various ontological views of species described above as the biological species concept, the SMRS concept, the evolutionary species concept, the phylogenetic species concept, or de Queiroz and Donoghue's synthetic concept held that species have a real existence in nature, though character evaluation might aid us epistemologically by allowing us to identify them. This ontology-based perspective can be constructively applied to the definition of areas, and when it is used, it means that areas do have a real existence in nature, which might be related to geological or climatic processes. The possession of individual, unique taxa or the distribution of geographic barriers can be used to aid identification but areas have a real existence regardless, which is the view that is advocated herein.

If areas have a real existence in nature, they also have a history. In this sense, they would be spatiotemporally restricted individuals. The history of an area would explain how it influenced the evolution of its biota. An area would be given cohesion by geological processes, such as plate tectonics or climate change, and these processes would influence the geographic barriers that delineate it and thus influence its biota.

Just as de Queiroz and Donoghue (1988, 1990) discussed in the case of species, there are (at least) two ways of viewing areas ontologically. One would be to view areas as entities having history, conforming to a shared pattern of common descent and monophyly, e.g., a single continent successively fragmenting into four smaller areas. This history of fragmentation could be expressed by means of a cladogram showing the shared pattern of common descent, geologically, and we would also expect it to correspond, in general, to patterns of evolution and speciation in the clades that occupied the craton ancestrally (see Fig. 6). In other words, these areas would be monophyletic and show a pattern of common descent. Then, they would be like taxa, corresponding to an ancestor and all of its descendants. This would be one way that areas could be approached in a vicariance biogeographic framework, if they were believed to have real ontological status. If this view holds true, then areas would be analogous to higher taxa and perhaps to evolutionary and phylogenetic species, and several authors have suggested the analogy between areas and taxa. In this case, the history of areas would be governed by a geological history that produced a pattern of monophyletic areas, which concomitantly influenced a pattern of vicariant differentiation in their taxa. Further, using this perspective, it would be reasonable to suggest that areas would be related through common descent. That is, they can show something analogous to a pattern of phylogenetic relationship in the context of the evolution of species and taxa.

However, this is not the only valid way to look at areas ontologically. As discussed in Chapter 6 there is abundant evidence that there are many types of geological processes. These, along with climatic change, can amalgamate regions, with small areas combining to make larger ones. In short, there is reason to believe that reticulation and area-joining is also an important part of Earth history. After such joining occurs, areas can later fragment and the process could occur repeatedly, which would imply that these areas have not evolved by a strict process of divergent common descent. Then, evolution of areas by reticulation would prevail. Formerly independent areas would be like independent lineages on a tree of life, and could become joined long after their ancestors had differentiated (see Fig. 10). When this happens, areas would not be related by a simple pattern of monophyletic descent from a common ancestor. Rather they might show something analogous to paraphyletic or polyphyletic taxa, with areas that do not even share a recent common history becoming conjoined at a later time. This means that areas can also show something analogous to tokogenetic relationships, viewed in the context of the evolution of species and taxa [see Hennig (1966) and de Queiroz and Donoghue (1988, 1990)].

As there is ample evidence that processes involving the collision of areas have occurred, and as it is easy to imagine how they can occur within a plate tectonic framework or as a result of climatic swings through time, it is reasonable to state that these have to be considered when we try to understand areas. Areas can also be related through a process of amalgamation, which would be directly analogous to interbreeding among organisms of a species. Populations of organisms within species that have separate histories of common descent can still come together for the purposes of reproduction; they can still interbreed. Similarly, areas that have been separated for long periods of time can later come together and merge.

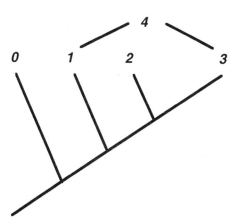

FIGURE 10. An area cladogram where there is evidence for an area hybridization event between areas 1 and 3 in area 4.

If we accept the arguments of de Queiroz and Donoghue (1988, 1990) then monophyly and interbreeding are both legitimate criteria in the definition of species. The same holds true for areas. Both amalgamation or geodispersal—equivalent to interbreeding—and fragmentation or vicariance—equivalent to common descent—describe important events in the history of an area. In short, areas are not analogous to taxa; they are analogous to species-level taxa. They have an existence by the principle of common descent as well as by their ability to interact and amalgamate with one another, which shows something analogous to both phylogenetic and tokogenetic relationships.

With this framework, we can also recognize that areas are individuals. They have some history during which they are spatiotemporally isolated. They have a birth, denoted by the emergence of geographic barriers that delineate the area, and they have a death, denoted by their effacement or removal. However, if we are to view areas as individuals, then when new barriers subsequently arise in the same or similar places to where they had been previously, it would be either necessary to hold that a new area formed or that the smaller-scale area maintained its identity even after it had amalgamated with another entity into a putatively larger unit. Further, we must remember that individual species have some cohesion. Areas also potentially have cohesion, which is provided by the persistence of the geographic barriers that surround them as well as by the continued absence of geographic barriers within their boundaries. These determine the distribution and evolution of the biota within any particular area, and this biota is subjected to a unique set of shared geological and climatic conditions, which cause it to maintain its independence from the biotas of other areas.

I have been arguing that areas have their own ontological status independent of the biota they contain, though geological and climatic events that affect them will surely affect their contained biota. Tied up with the definition of areas is the notion of a shared, unique geology and climate. Historical biogeographic patterns of distribution and evolution are produced primarily by the rise and fall of geographic barriers, which cause vicariance and geodispersal, so it is these processes that should be of most interest to historical biogeographers.

7.5. The Epistemology of Areas

Discussions about the nature or ontology of areas are important, but we must also consider the epistemology of areas, or how we identify them. Few, if any, rigorous ways to identify areas of endemism have been developed, though such methods would be very useful (Harold and Mooi, 1994; Morrone, 1994).

A common procedure for identifying an area of endemism involves identifying one or more unique species within it, which is an epistemological procedure that emphasizes biological data. In the case of a single unique species, are the boundaries of the area the same as the geographic distribution of the species? Determining the margins of the geographic distribution of a species may be problematic because these boundaries can be very amorphous, moving back and forth as a single organism expands or contracts its range daily or seasonally. (This would be impossible to determine in the fossil record.) Further, a species is unlikely to occupy every square inch of a region; there will be gaps within its overall range. In addition, should only breeding individuals of a species be considered or should all individuals be included in a tabulation of range (again determining this is very problematic in the fossil record)? How should these problems be treated?

The issue becomes even more complicated when an area contains several species. The margins of the geographic ranges of these species are unlikely to overlap completely (Axelius, 1991; Harold and Mooi, 1994; Hovenkamp, 1997), and again there is the question of gaps within the range of any particular species. Further, some of the ranges may differ significantly in size, such that the range of one species is only a small subset of the range of another. What is the boundary of the area in this case? It would seem that to determine the boundaries of an area some statement about the geographic ranges of its contained taxa might be necessary. In this case, ranges that differ subtly might be treated as homologous, just as in the case of differences in anatomical structures that are still believed to be shared owing to common descent.

From a biogeographic standpoint, any slight difference in geographic range might be treated as noise in an overlaid signal related to historical biogeographic events and the boundaries of a biogeographic area. This has been expressed in the following way:

> An area of endemism is an area of nonrandom distributional congruence among different taxa. It is identified by the congruent distributional boundaries of two or more species, where congruence does not demand complete agreement on those limits at all possible scales of mapping, but relatively extensive sympatry is a prerequisite. (Morrone, 1994, p. 438).

Axelius (1991) suggested a different approach; he concluded that since the geographic distributions of species often partially overlap, it is best to consider the regions of overlap and the regions of nonoverlap as distinct areas. However, this protocol seems too rigid. Any overlap in geographic range might mean that these species have been influenced by the same geological or climatic processes and thus might provide some biogeographically relevant information. Thus, this proposal potentially obscures resolution in biogeographic analysis and is not recommended.

To a certain point, when defining areas it is important not to lose the forest for the trees. Subtle differences in geographic range of taxa need not be considered when determining the boundaries of an area because one will miss the big picture. However, it is important to recognize that having quantitative techniques to define areas would be highly beneficial.

Morrone (1994) has suggested one possible method where square-shaped quadrats are drawn on a map, and then a data matrix is constructed with the quadrats on one side of the matrix and the species on the other. A species is recorded in the data matrix as either present "1" or absent "0" in a quadrat. This matrix can then be analyzed using a parsimony-based algorithm such as PAUP 4.0 (Swofford, 1998). The output will be a cladogram depicting the relationships of different quadrats. Quadrats that share species will cluster and large clusters can be split off as distinct areas. Hovenkamp (1997) propounded a similar technique.

Although an important step in the right direction, there are some problems with Morrone's (1994) method. First, the analysis of a matrix may produce several clusters of quadrats on the cladogram. The choice of which hierarchical level is the appropriate cut-off to delimit areas becomes exceedingly arbitrary. Several different clades can be identified. At what level is a clade of quadrats equivalent to a biogeographic area? Further, the size of the quadrats that is chosen has *a priori* a very strong influence on the results, and as was the case with clade size, the choice of quadrat size is arbitrary. Therefore, although there is too much about this method that is arbitrary, still a biologically based approach to defining areas will be successful, as long as one recognizes that there is some ambiguity in defining geographic ranges and, further, if one appreciates that these ambiguities might influence subsequent results.

In addition to biologically based procedures, there could also be an epistemological procedure for discovering and delineating areas that involves geological information. The boundaries of an area might be defined by major geographic or physiographic barriers that are related to geological features or climatic differences. In this case, areas would not initially be equated with areas of endemism, but they might be expected to show a strong correspondence to lines mapping the geographic distributions of species, such that in the end geologically defined areas would be equivalent to areas of endemism. For instance, for terrestrial organisms the biogeographic area of interest might be defined as a land mass surrounded on all sides by an ocean. The size of that land mass will vary slightly with, e.g., tides and storms, but the boundaries of the area could be treated as equivalent to the watermark at low tide. The geographic ranges of terrestrial organisms on this land mass will generally not extend beyond the geographically defined boundaries of the area.

Major mountain ranges could also be used to define areas within continents. The mountain ranges could be the boundaries that separate one continental area from another and act as geographic barriers to the free movement of terrestrial organisms. For freshwater organisms such as fish, areas might be the entire drainage of a river system, the bounds of which reflect underlying topography and climate (e.g., Rosen, 1978; Wiley and Mayden, 1985; Mayden, 1988).

For nearshore marine organisms, boundaries could be defined as the continental shelves of a land mass on one side and the continental slope on the other. Much of the marine invertebrate fossil record is a chronicle of life in shallow seaways that once covered the present-day continents. There were often basins within these continents, topographic depressions surrounded by emergent or elevated swells of land, in which there were continuous marine deposits, so exceptional fossils were preserved. For example, in the Paleozoic

FIGURE 11. A reconstruction of eastern North America (ENA) showing the approximate outer margins of the major tectonic basins in this region during the Middle Devonian. 1 = the Appalachian Basin, 2 = the Illinois Basin, and 3 = the Michigan Basin. From Lieberman and Eldredge (1996), used with permission.

era eastern North America contained three major tectonic basins (Fig. 11), with geometries that are easy to trace. Further, the basins are surrounded by elevated arches, which are also easy to trace, and, depending on the sea level, these arches would have served to isolate the marine faunas present in those basins from one another. In this case, the biogeographic areas could have been defined as corresponding to the limits of these basins. In both ancient and modern oceans, the boundaries of areas could be defined on the basis of either surrounding continental masses or current systems.

If one believes that the ontological status of an area is derived from the fact that it has a unique geological and/or climatic history, then the geologically and geographically based definition of areas might be more accurate. However, there may be cases where some geographic barriers within regions that appear to be insignificant play an important role in influencing the evolution and distribution of biotas. Other times, geographic barriers that might be expected to be very prominent have little influence on biogeographic patterns. Further, at the large scale, geological information may work for defining broad areas of endemism, but there may be many such smaller areas within the geologically defined larger area.

In the face of the foregoing facts, some combination of geological and biological information might be the best possible approach to defining areas worthy of biological study, though this is somewhat nebulous. Information about geographic barriers as well as the geographic ranges of species would be used to define areas. Barriers that corresponded to the geographic ranges of numerous species might be interpreted as the boundaries of an area. By contrast, barriers that seemed to be unrelated to the geographic distributions of species could be ignored. In the end, the epistemological delineation of areas seems mildly unsatisfactory, like the delineation of species-level taxa. Maybe this is just the nature of the problem of defining areas and it can never be more thoroughly resolved, just as there is no means for always identifying a species-level taxon in any given case. However, just as the method of identifying a species as the smallest putatively interbreeding group of organisms that shares some unique morphological features generally works (e.g., Jackson and Cheetham, 1990), so too does the method of defining areas by mapping out the geographic distributions of species and looking for many lines that are similar and overlap, particularly those corresponding to ancient or modern geographic barriers.

It should be noted, though it is perhaps not surprising, that in some studies any of several areas might be used. The sizes of the areas chosen often depend on the desired limits of resolution and the goals of the study. Which areas are used and why should be stated up front because the choice of areas might possibly influence the results of the study. For example, we can imagine that six areas of endemism actually existed within a broad region: A, B, C, D, E, and F. We assume that the actual geological history of this broad region reflected the following sequence of fragmentation [A, (B, C), (D, E, F)]. Let us imagine further that because some taxa were distributed in both C and D a biogeographer treated them as a single area of endemism, J. A very different pattern of area relationship might then be uncovered by this analysis.

7.6. Conclusions

In summary, there are several interesting issues that emerge when considering the ontology and epistemology of areas. First of all, ontologically, areas show many similarities to species (other parallels between biogeography and systematics will be explored in the next chapter). Therefore, a possible resolution to the debate about the nature of species offered by de Queiroz and Donoghue (1988, 1990) seems to provide an excellent means for thinking about the ontology of areas. Areas can be viewed as entities that display a pattern of common descent, with larger areas breaking up into smaller ones. This is akin to a pattern of phylogenetic relationship, and it is the framework typically considered in the vicariance school of biogeography. However, areas can also have something like the reticulate tokogenetic relationships seen within or across traditionally defined species, which can show patterns of interbreeding among populations that have had long, independent histories. In this case, different areas are of course not reproducing with one another, but they would be colliding and amalgamating to form larger areas. This perspective has often been ignored by vicariance biogeographers, but it is an equally valid one and can be produced by the process of geodispersal discussed in Chapter 6. Epistemologically, the definition of areas still lacks rigor, much like the definition of species. Perhaps more quantitative methods for defining areas will be developed, but it is hard to imagine how this can be done without there being some degree of arbitrariness in the analysis.

Chapter 8

Biogeography and the Comparative Method

8.1. Introduction

Advances in biogeography and paleobiogeography are very much linked to the expanding prominence of phylogenetic systematics as a research program in evolutionary biology and paleobiology. Phylogenetic systematics, as part of the comparative method in biology, emphasizes the evolutionary relationships among groups of organisms as an important aspect for testing hypotheses about the nature of the evolutionary process. For instance, to understand whether or not a particular trait is an adaptation and how selection pressures may have influenced its development, we need to know its primitive condition and how it changed with evolutionary events in the relevant group. In other words, we have to have some understanding of phylogenetic relationships within the group. As another example, if we want to look at how climate change has influenced the evolution of a particular group, we have to know the sequence of branching events within it. The relevance of phylogenetic systematics to a research program in evolutionary biology is part of a broader framework for testing hypotheses in geology and biology. In the historical sciences we are confronted with patterns, such as the movement of glaciers, the movement of continents, and the evolution of terrestrial organisms, which we wish to explain by invoking one or more processes. In each of these cases, before historical scientists such as geologists or biologists invoke a process to explain a pattern, they have to know what that pattern actually was.

Phylogenetic analysis was formulated as part of an effort to develop a rigorous means of determining the evolutionary relationships among groups of organisms. These evolutionary patterns can then be used to make statements about the evolutionary process. In short, "the most important connection between the two areas (pattern and process) involves the comparison of the patterns of both intrinsic and extrinsic features of organisms predicted from theories of process with those actually 'found' in nature" (Eldredge and Cracraft, 1980, p. 4).

Phylogenetic analysis has been especially critical for evolutionary biologists interested in analyzing macroevolutionary phenomena (Brooks and McLennan, 1991). It can be conducted at any level of the genealogical hierarchy, but at the higher levels, where traditional experimental approaches are not possible, its methods are the most powerful available to study evolution rigorously. For example, analysis of the evolution of clades cannot be studied experimentally (except through numerical simulations) and thus is most accessible through phylogenetic studies.

Phylogenetic analysis stands out because it introduced rigor to the study of macroevolution, but it is not the be all and end all of macroevolutionary studies. Other techniques to ascertain evolutionary relationships may emerge in the future, and there is exciting work going on now to find new ways of analyzing evolutionary relationships. However, as discussed in Eldredge and Cracraft (1980), Wiley (1981), and Brooks and McLennan (1991), phylogenetic analysis was a major advance as prior to its emergence, hypotheses concerning the large-scale pattern of evolution were presented as treelike scenarios. Further, many parts of these scenarios were often unresolved polytomies, which were agnostic statements about detailed aspects of evolutionary relationships. Most crucially, little evidence was presented to support either aspects of the scenario or the scenario in toto, such that it became hard to evaluate different scenarios or to determine why a particular one was correct. When evaluations were done, they often deteriorated into invocations and arguments about the scenario builders' status in the field. Further, there was no sense that uniquely held characters, or primitive characters, might not convey information about evolutionary relationships.

Part of the difficulty was that a rational philosophical approach had not been promulgated, but perhaps a more crucial problem was an analytical one. As demonstrated by Bowler (1996), evolutionary relationships and the origins of various higher taxa have been debated throughout the history of evolutionary biology. In the 19th century and for most of the 20th century these debates could not be resolved because some character systems suggested one set of evolutionary relationships and other systems suggested other schemes. Moreover, when evolutionary biologists were interested in the evolutionary relationships of many species at once, it became very difficult because it is hard to consider many different sets of characters from many species at the same time without some kind of advanced computing device.

The development of phylogenetic analysis, spurred by the publication of the English translation of Hennig's (1966) magnum opus, lead systematists to realize that only certain types of characters, shared derived characters or synapomorphies, should be used as evidence of evolutionary relationships. However, the analytical problem still remained. Determining these relationships among groups of organisms often entailed many competing sets of characters, some of which indicated that two species A and B were most closely related to a third species C, whereas some characters indicated that species B was most closely related to another species D. Thus the development first of personal computers and later of computer algorithms was of enormous importance to the field.

Computer algorithms, such as Hennig 86 (Farris, 1988) and especially PAUP (Swofford, 1993, 1998) can be used to choose among competing character data from several species to determine which pattern of evolutionary relationship is best supported by the available data. Another important computer package that facilitates the analysis of character data in phylogenetic studies is MacClade (Maddison and Maddison, 1992). These algorithms not only provide a way of analyzing character data to come up with an evolutionary tree or cladogram, but also made it possible to quantify the support for a given cladogram (or even the support for parts of that cladogram).

Phylogenetic systematics was neither the first nor the only approach to analyzing patterns of relationships among organisms. The initial advances toward analytical rigor in systematic research were undertaken by the pheneticists (see Eldredge and Cracraft, 1980). Unlike in phylogenetic systematics, there was no one algorithmic approach preferred by all pheneticists (Hull, 1988), and their studies over the years have incorporated a range of multivariate statistical techniques. One critical distinction between pheneticists and phylogenetic systematists is in the type of characters they used as data to make statements about the relationships among organisms. The latter argued that characters unique to single taxa, or autapomorphies, and characters that were primitive within a group of taxa, or plesiomorphies, should not be used to determine evolutionary relationships because they provided no information about patterns of common descent. By contrast, the former, although they were not always interested in deriving statements about the evolutionary relatedness of taxa but rather about natural classifications of these taxa, did use both autapomorphies and symplesiomorphies in their classifications.

Although the development of a rigorous analytical approach in systematics was very important, it did not by itself facilitate major advances in the field. For philosophical reasons discussed in Eldredge and Cracraft (1980), Nelson and Platnick (1981), Wiley (1981), to cite only a few, the selection of appropriate character types was also a critical factor.

As shall become apparent in the next chapter, a similar phenomenon emerged in the area of biogeographic research. The development of rigorous analytical techniques for deducing biogeographic patterns was extremely important for the discipline of biogeography. However, these techniques alone do not suffice, and it was the recognition that only certain types of biogeographic data should be used that greatly facilitated advances in this area.

8.2. Phylogenetics and Biogeography

All of the advances described above were crucial in making phylogenetic systematics a rigorous part of the science of evolutionary biology. Now how does this tie into the field of biogeography, and what is the relationship of phylogenetic systematics to biogeography? Biogeography focuses on the distribution of organisms and the relationship of their distribution either to their evolution or to their ecology. The study of distributions of entities in the economic hierarchy is a subdiscipline of ecological biogeography, and the study of the distributions of entities in the genealogical hierarchy is a subdiscipline of historical biogeography. Historical biogeographic patterns are conjoined to evolutionary patterns so there must be an important role in historical biogeography for a technique that assesses evolutionary patterns, namely, phylogenetic systematics. If we hope to look at how the geographic distributions of groups of species have changed as they have evolved, then surely we need to know how these groups of species evolved, and any rigorous technique that can be brought to bear must be important.

However, the link between phylogenetic systematics and biogeography runs even deeper than that. The key goal of any study in historical biogeography is to ascertain whether or not groups of species responded in unison to a set of geological or climatic changes (Brooks *et al.*, 1981; Brooks, 1985, 1990; Wiley, 1981, 1988*a,b*; Wiley *et al.*, 1991; Brooks and McLennan, 1991). If this was the case, then the sequence of geological or climatic changes influenced a sequence of evolutionary events in several species, and there were concomitant changes in the geographic ranges of species in these clades. What we should see, as an ideal end result, is that biotas evolved in unison. That is, as historical biogeographers, the pattern we are trying to recover is whether or not biotas evolved, in terms of the geographic distributions of their component species, in unison.

The metaphor of the tree or cladogram, so important for evolutionary biology, comes in here, for if biotas did evolve in unison, or largely in unison, then another way of expressing this pattern would be as a tree relating different areas. Relationships among areas would provide information about the relative time that those areas either separated, owing to such processes as continental rifting, or amalgamated, owing to such processes as continental collision. Respectively, species from these two types of closely related areas would recently have either diverged (owing to the formation of geographic barriers and subsequent vicariance through allopatric speciation) or expanded their geographic ranges (owing to the fall of barriers and geodispersal), such that their distributions became overlapping. The best way to represent both evolutionary and biogeographic relationships is as a tree or cladogram, keeping in mind that evolution occurs by the historical process of common descent and that plate tectonics is also a process that causes the continents and their interactions to have a history. How we construct a tree describing reticulating events will be described in Chapter 9.

Means of evaluating evolutionary and biogeographic relationships also have similarities. Phylogenetic analysis is a discovery technique that looks at shared patterns of characters to infer how groups of species are related evolutionarily. These characters are attributes of organisms that can be considered evidence bearing on hypotheses of evolutionary relationship. Characters could be discrete features of an organism's morphology, such as the presence or absence of certain bones, or DNA sequences, or quantitative measurements of certain traits. If two groups share many uniquely derived characters with one another and no derived characters with a third group, then it is assumed that they share a more recent common history from which the third group is excluded. Unfortunately, the situation of how shared derived characters are distributed among different species is rarely clear-cut. Often, some characters suggest one pattern of evolutionary relationship while other characters suggest another. One must remember that this was precisely the problem that led to the original dissatisfaction with approaches in evolutionary biology that sought to ascertain the large-scale evolutionary history of life (see Bowler, 1996). How can we proceed from here?

In phylogenetic analysis competing character systems are evaluated to determine the best-supported pattern of evolutionary relationship using the principle of parsimony (Gaffney, 1979; Eldredge and Cracraft, 1980; Wiley, 1981). In this approach, the relationship best supported by the available data is equivalent to the cladogram that requires the fewest number of steps and thus the fewest independent, multiple origins of characters. This cladogram, by invoking the fewest number of independent origins of characters, also makes the fewest *ad hoc* statements about the character data; i.e., characters are treated as real, and the shared possession of characters is *a priori* evidence for evolutionary relationship. *Ad hoc* statements about characters would involve the supposition, prior to a rigorous analysis, that although a character looked the same in different species, it evolved independently.

The principle of parsimony is really just an evaluation procedure to be employed when one is confronted with conflicting data. There are philosophical justifications for using this method, discussed by Wiley (1975, 1981), Platnick and Gaffney (1977, 1978a,b), Gaffney (1979), Eldredge and Cracraft (1980), Sober (1988), Wiley *et al.* (1991), and Brooks and McLennan (1991). However, there is also the common sense justification that the principle involves looking to see which pattern is supported by most of the data. Part of the rationale for accepting the principle of parsimony in the first place is that there is strong evidence for the existence of a hierarchy of life with groups of organisms related in ever-broadening sets or branches on a tree. This hierarchical pattern was one of the reasons the idea of evolution by common descent was formulated. [see, e.g., Lamarck (1809), Wallace (1855), and Darwin (1859)]. The hierarchical pattern presumes that monkeys look more like humans than whales do, and therefore monkeys and humans share a more recent common ancestor than either shares with whales. Associated with this shared pattern of common descent is the joint possession of a greater proportion of shared derived morphological features because of their longer-shared common evolutionary histories.

To perform a biogeographic analysis, and make a determination of area relationships, we need data or certain kinds of evidence. In phylogenetic analysis this evidence involves characters that these are arranged in hierarchies with the evolution of life expressed as a pattern of common descent. Is there anything equivalent in the field of historical biogeography? What kind of data or evidence bears on biogeographic studies? First, there are the patterns discussed in Chapter 4, and it has long been recognized that different regions are characterized by different complements of species, and that the most closely related species (evolutionarily) are typically found near one another. These twin phenomena constituted some of the actual biogeographic patterns that first convinced scientists of the need for the field of biogeography, and then, later, of the importance of the field for the study of evolution and the movements of continents. The existence of these patterns constitutes some gestalt justification for the existence of a set of biogeographic data that can be evaluated to study the history of areas. This is just as the pattern of hierarchical degrees of resemblance of organisms constituted some gestalt justification for evolution by common descent and the existence of a set of character data that can be evaluated to study evolutionary relationships.

As discussed earlier, historical biogeography is the discipline that looks at how groups of organisms have evolved and how their geographic distributions have changed in relation to geological or climatic events. Thus, the types of data that bear on hypotheses about the historical biogeography of particular regions relate to how groups of organisms are distributed and how they have evolved. (Detailed discussions of how such abstract concepts as evolution and geographic distributions can be converted into data will be found in Chapter 9.) In phylogenetic analysis, the arbiter among competing hypotheses suggested by different character systems, i.e., incongruence among characters, is parsimony. The analogous problem in biogeography is what to do when one group suggests one biogeographic pattern and another group suggests another. This is a critical area in biogeography, which will be considered in greater detail in Chapter 9. However, for now, it is only necessary to assume that cladograms for several groups of organisms and information about the geographic distribution of these organisms provide evidence, i.e., characters or attributes, that can be used in a biogeographic study. If we can track how a set of organisms evolved, and how their geographic distributions changed as they evolved, we can look at the history of the formation of barriers that produced speciation via vicariance and at the history of the elimination of barriers that produced range expansion via geodispersal.

8.3. Tracing Biogeographic Distributions Group by Group: the Problem of Ancestors

Patterns of evolution and changes in geographic distribution at this stage have to be traced group by group. In other words, before we worry about what happens when each of two groups suggests a different biogeographic pattern we must first consider what we mean by biogeographic patterns in a single group. The best way to think about this is to imagine an evolutionary tree or cladogram relating a set of species, say, four trilobite species (Fig. 12). If we want to look at how this group evolved related to changes in geographic distribution then we have to know something about that distribution. Let us substitute the areas these species occur in for the species names (Fig. 13). Now we know that the closest relative of a species from the Michigan Basin is found in the Illinois Basin (see Fig. 11 for location of different basins in eastern North America). However, this is not enough information to reconstruct how geographic distributions changed as this group evolved, because we also have to know something about where the ancestor of these two species occurred. How can we derive this information? A broader problem would be to consider several species in a clade and determine the clade's ancestral geographic range.

A traditional paleontological approach would be to look for the oldest species in the group, see where it occurred, and then assume that this area was the group's ancestral homeland. However, this may be both naive and inaccurate, for it includes an assumption that that the oldest species in the group is the direct ancestor of all the other species, which is not always a reasonable one. Why might it not be reasonable? To provide an answer we have to refer again to the nature of cladistic analysis.

In a cladistic context it is difficult to deal with ancestors (though this is not a compelling enough reason to abandon their usage for it relates to methodological principles rather than to the nature of the history of life). If

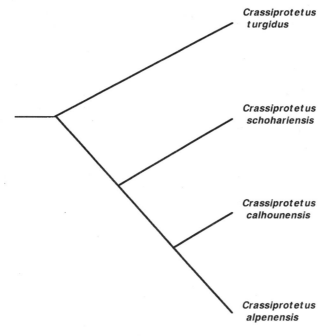

FIGURE 12. A phylogeny relating four species of the trilobite genus *Crassiproetus*.

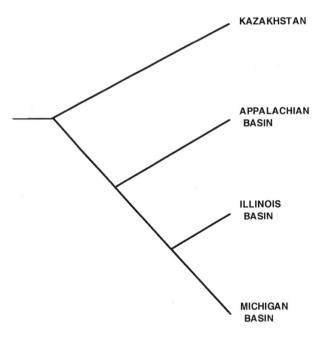

FIGURE 13. An area cladogram showing the distribution of the species of *Crassiproetus* substituted for the taxon name.

there were a cladogram for this group, initially no species would be treated as the direct ancestor of any other species. Rather, the cladogram would be a statement about a hierarchy of shared common ancestry, but at each node, situated at a bifurcation, there would be a hypothetical ancestor rather than a real one (Eldredge and Cracraft, 1980). Methodologically, in order for an actual species to be a direct ancestor of one or more other species it must have an exactly intermediate distribution of characters, while not possessing any unique characters. Of course, there might be ancillary evidence that could be brought to bear on a hypothesis of direct ancestry. For example, we might imagine that the two trilobite species *Crassiproetus calhounensis* and *C. alpenensis* overlap stratigraphically, and that the former has characters that are so close to those of the latter, without possessing any characters unique to it, that it would seem to represent a good ancestor.

In such a case, reconstructing the area in which the ancestor of *C. calhounensis* and *C. alpenensis* was distributed would be very straightforward: it would be equivalent to the range of *C. calhounensis*, within the Illinois Basin. Then, sometime during the history of the genus *Crassiproetus*, one species moved from the Illinois to the Michigan Basin and there was a speciation event.

The problem of identifying ancestors is a very important and troubling one in evolution, paleontology, and biogeography. In most cases it is not possible to recover the direct ancestor of any descendant species (Engelmann and

Wiley, 1977). Either there is not sufficient stratigraphic overlap or the postulated directly ancestral species does not possess the exact mix of characters necessary to dignify it with the title "ancestor."

There is actually a reason that we might not expect to recover direct ancestors, which has to do with the use of stratigraphic overlap as a proxy for ancestral status, and it relates to the nature of the evolutionary process. If punctuated equilibria (Eldredge and Gould, 1972) is an adequate description of much of evolution, then speciation occurs owing to the isolation of small populations of a large species at the periphery of that species range. Even after speciation, these small populations, which occur in narrow geographic regions, are very unlikely to be preserved in the fossil record. Only if speciation is successful and the species reinvades into the original area without extirpating the ancestral species will a direct case of ancestral–descendant speciation be captured in the fossil record. More likely, because the very conditions that promote evolutionary change retard the ability to recover newly divergent species in the fossil record, we would expect that at the finest levels, speciation within clades, it would be stratigraphically gappy. This pattern is in fact a direct prediction of the punctuated equilibrium paradigm, violated only when stratigraphic completeness reaches unusually high levels (e.g., Williamson, 1981).

If the exact ancestral species is not known, how can we infer the geographic distribution of the ancestor of what are now two, or more, species? Fortunately, this is a problem that has been considered from a different standpoint. In the fields of systematics in particular, and evolutionary biology in general, it is often important to be able to evaluate how and why an anatomical or a molecular character evolved in a group over time. In order to do this, we need to know the ancestral condition of that character in that group before it diverged evolutionarily (Harvey and Pagel, 1991; Brooks and McLennan, 1991; Maddison and Maddison, 1992). Numerous techniques have been developed, detailed in Maddison and Maddison (1992) and elsewhere, that describe how this can be done, and the same kinds of techniques can be applied to problems in biogeography; i.e., they can be used to make inferences about the ancestral geographic state of a group.

8.4. Quantitative Approaches to Reconstruct the Historical Biogeography of Individual Clades

The application of these techniques to problems in the biogeography of single groups was specifically discussed in Bremer (1992, 1995), Ronquist (1994, 1995, 1997), and Lieberman and Eldredge (1996). Bremer's discussions and the first two of Ronquist's focus on the problem of how we can reconstruct the geographic distribution of the single common ancestor of an entire clade. Their approaches were not originally designed to study the distributions for each of the various ancestral states that make up a clade. We considered the latter problem (Lieberman and Eldredge, 1996) as did Ronquist (1997).

Bremer (1992) defined the problem of identifying the distribution of the ancestor of the group and equated it to that of finding an ancestral area. Before we can determine an ancestral area, we need some understanding of both the geographic distribution of taxa in the clade and the evolutionary patterns in that particular group. The complexities of the first problem were discussed in Chapter 7, but for present purposes let us assume that the geographic ranges of each species within the clade are well characterized. After we have this information there are two ways of approaching the problem (Bremer, 1992), and these are described in what follows.

8.4.1. Areas as Multistate Characters: Parsimony-Based Approaches

In the first approach, which is based on the work of Mickevich (1981), we can think of each area as representing the states of a single, multistate character. An analogy from phylogenetic analysis explains how this would work. We might be interested in deducing the evolutionary relationships among a small group of dinosaurs that differed in the number of spines on their backs; say, some species had 5 spines, some had 7, some had 13, and some had 22 spines. Further let us say that we knew that 5 spines was the primitive condition in this clade. Then spines in these dinosaurs could be coded as a character with four states: 0 = five spines; 1 = seven spines; 2 = 13 spines; and 3 = 22 spines (Fig. 14). This character, in combination with others, could be used to perform a phylogenetic analysis to assess how these species were related. Returning to biogeography, in the case of geographic areas, each of the states of a multistate character would represent a species' occurrence in a particular area or geographic region: e.g., 0 = Australia, 1 = South America, 2 = Antarctica, and 3 = South Africa. We imagine that the aforementioned spiny dinosaurs were known as fossils from each of these regions. These biogeographic character states could then be placed on the evolutionary tree, substituting them for the original species names (Fig. 15). In this sense, geographic distribution is like any other character of a species.

Once we know the condition of the terminal taxa, with the help of a computer package such as MacClade (Maddison and Maddison, 1992), it is relatively straightforward to determine the conditions at the nodes representing the hypothetical ancestors. This approach, originally proposed by Mickevich (1981) should be distinguished from that of Bremer (1992, 1995) and Ronquist (1994, 1995) because it takes the biogeographic states of all nodes or ancestors within a clade into account, rather than just that of the last common ancestor. There are different algorithms to determine the state of the ancestor, but those used most consistently are parsimony-based and they assume that character changes occur in the minimum number of steps. Although this assumption can be problematic, the alternative, rejecting parsimony, is even more problematic. It would mean that any number of

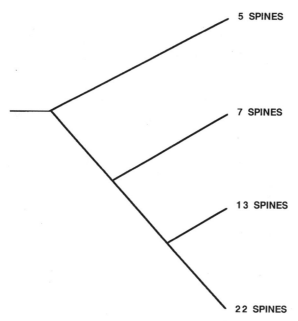

FIGURE 14. A cladogram relating four hypothetical species of dinosaurs, where the number of spines each species has on its back has been substituted for the taxon name.

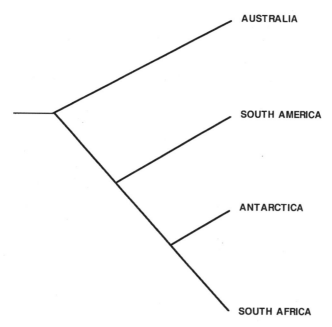

FIGURE 15. An area cladogram with the geographic distribution of spiny dinosaurs substituted for the number of spines.

transitions could occur and thus any possible optimization of ancestral states is theoretically possible. At present, although parsimony-based algorithms are the most frequently utilized, others are being developed based on maximum-likelihood approaches [see Schultz *et al.* (1996) and Schluter *et al.* (1997)]. These methods have significant strengths, including the fact that they allow hypotheses about ancestral character states to be placed within a rigorous, statistical framework, but they also have significant weaknesses. Chief among these weaknesses is the fact that their use requires some knowledge of probabilities of change. Often, we cannot honestly state what those probabilities are between one character state and another, regardless of whether the states are features of morphology or geographic distributions; yet statements about such probabilities would clearly influence the results. Since the application of such maximum-likelihood approaches to character optimization in general, and the optimization of biogeographic characters in particular, is only in its infancy, and at the moment appears problematic, it will not be discussed any further here.

In regard to parsimony-based algorithms, there is a range of potential choices that can be utilized. For example, there are algorithms that assume that characters transform in a specified order, let us say from state 0 to state 1 to state 2. These are so called ordered or additive optimization algorithms [see Maddison and Maddison (1992) for discussion]. This type of algorithm might be worth using if we were sure that the pattern of divergence across geographic lines among, say, the spiny dinosaurs across geographic lines always proceeded from Australia to South America to Antarctica, etc. Sometimes we know this beforehand, but, often this is the very type of biogeographic pattern we are trying to determine, and we have no way of knowing, *a priori*, the pattern of relationship among different geographic areas. Therefore, the use of ordered algorithms is not generally recommended.

There are other types of parsimony algorithms that make certain assumptions that may be valid in some types of evolutionary studies, e.g., how characters evolve, but which may not be realistic in biogeographic studies. Included among these are the algorithms that assume *a priori* irreversible patterns of transformation in characters (Camin–Sokal parsimony) or that the same character does not evolve independently (Dollo parsimony). Neither of these translates freely to biogeographic studies, for the former assumes that geographic differentiation across a clade must always occur in a specified manner, and the latter assumes that the same biogeographic state (geographic distribution) is never reoccupied by another species. There may be specific instances in which a particular geography guarantees that one or the other of these scenarios prevails but, again, this is usually the type of phenomenon that we are trying to investigate. We want to know the pattern of transitions between areas as a group evolved, and it seems unwise to invoke *a priori* specific assumptions about how geographic differentiation occurred. [Ronquist (1994, 1995) leveled a further criticism of irreversible-parsimony approaches to reconstructing ancestral biogeographic states, albeit in a slightly different context.)

8.4.2. Using Unordered Parsimony to Study Biogeography

Another parsimony-based algorithm that can be used to optimize multi-state characters (geographic states) to ancestral nodes is that of Fitch (1971). This algorithm assumes that multistate characters are unordered. In the case of morphological characters such as the spines of dinosaurs, accepting this approach means that during the evolutionary history of a group any character can transform into any other character. Specifically, it might mean that 5 spines is as likely to transform into 7 spines as it is into 22 spines. Often, when we study the evolution of morphological characters, this is the most valid type of approach. We frequently have no *a priori* way of determining if the 5-spine condition is any more likely to transform to the 7-spine condition than to the 22-spined condition.

This is even more true in biogeographic studies. We may not have any idea *a priori* which way transformations occur between different geographic regions as a clade evolves. Rather, that is precisely what we are trying to investigate. Because the Fitch (1971) algorithm is suitably agnostic it seems to be a good first-order discovery procedure. This approach to determining patterns of geographic change within individual clades was the one we utilized, but we introduced certain modifications into the algorithm to make it more effective in a biogeographic context (Lieberman and Eldredge, 1996; Lieberman, 1997). Specifically, the original Fitch algorithm describes a set of protocols to determine the most parsimonious solution for ancestral nodes of a cladogram as follows:

Phase 1. (The Preliminary Phase): The geographic states at this point should be at the tips of the species tree, having replaced the names of the taxa; the method then proceeds from the descendants (the terminal taxa) to the ancestors, starting with those that are immediately ancestral to the terminal taxa. The ancestor is assigned geographic states, based on those that are present in its immediate descendants. If the descendants share no geographic states in common, then all of the geographic states are utilized, which is referred to as taking the union of these states. If they share some or all of their states in common, then only the states held in common are utilized in the ancestral reconstruction, which is known as taking the intersection of these states. The method then proceeds backward toward the base of the tree, which is the last common ancestor of the entire clade. This preliminary phase is followed by a six-step second phase that is employed in the following sequence and can be described as a sort of flowchart. Unlike the first phase, which proceeded from top to bottom (descendants to ancestors), the second phase proceeds from bottom to top, or from basal ancestors to more derived ancestors and descendants.

Phase 2: (*i*) if an ancestral node has all of the geographic occurrences present in its immediate ancestor go to *ii*, otherwise go to *iii*; (*ii*) delete all of the geographic occurrences not present in its immediate ancestor and go to *vi*; (*iii*) if in the preliminary phase the union of the states was taken, go to *iv*, otherwise go to *v*; (*iv*) add to the ancestral node any states that are present in its immediate ancestor that were not present after the preliminary phase and go to *vi*; (*v*) add to the ancestral node any states that are present in the final set of the immediate ancestor and in at least one of the immediately descendant sets and go to *vi*; (*vi*) the final phase is complete; go to the next ancestral node up the tree and return to *i*. This method was explicated in greater detail by Fitch (1971), and in practice, it would work as shown in Fig. 20.

8.4.3. Modifying Fitch Parsimony for Use in Historical Biogeographic Studies

As traditionally utilized, Fitch parsimony only allows one state to map to a node rather than the several different states that might emerge in phase 2. Fitch (1971) also described how this is implemented. He employed this additional step because he did not always reach parsimonious solutions after phase 2, and some of the assumptions about multiple states for a character at a node may not have been realistic in the molecular sequence data he was describing. However, whether or not we should utilize this additional step when we apply this optimization algorithm to biogeography is a different matter, and this has important implications for our use of it in biogeographic studies.

In the evolution of a clade there may be cases where we would believe that an ancestor should be more broadly distributed than its descendants, particularly if speciation occurs via vicariance. Then, if the optimization algorithm is realistic, ancestral nodes should at times have optimizations that indicate that the ancestor was distributed in several areas. However, with traditional Fitch parsimony, this is impossible. This problem was summarized by Ronquist (1997, p. 196), "(m)ethods such as Fitch optimization normally restrict polymorphism (widespread species) to terminals, and ancestors are exclusively reconstructed as monomorphic (occurring in single areas)."

It is clear that this traditional Fitch approach is problematic for several reasons when applied to biogeographic studies. First, it causes problems if we are trying to synthesize biogeographic patterns from several clades (described in Chapter 9) as it leads to several trees for each cladogram, each representing different solutions for optimizing ancestral biogeographic states. Further, intuitively, an approach that allows multiple states to map to a single node makes more sense because each state is one of several potential regions that an ancestor may have occupied before it speciated or before it expanded its range. For this reason, the approach shown herein, and the one that we used (Lieberman and Eldredge, 1996; Lieberman, 1997) allows multiple biogeographic states to map to a node at the same time (i.e., the broadest possible interpretation of the Fitch algorithm that would terminate after phase 2). With this special implementation of this algorithm, we have a method for reconstructing the geographic states of ancestral nodes that does not make burdensome assumptions about the biogeographic process. As it allows transitions between different biogeographic regions to occur in any direction, it is realistic in terms of the way the speciation process works, and moreover, it is analytically tractable. When used this way, the algorithm does not always guarantee most-parsimonious reconstructions, but the pathways to such a reconstruction are contained within the multiple states at nodes.

Although the Fitch (1971) unordered approach to character optimization has definite strengths, there may be times when several different groups in certain areas have already been surveyed so that we have a good idea about how geographic distributions changed as these groups evolved and to what extent different regions share geographic history. In such cases, it may be possible or even desirable to use less agnostic approaches which do not make it equally likely for transitions to occur between different geographic regions. Then, ordered character approaches could be used.

The advantage of ordered over unordered approaches is that the former, because they allow transitions between any regions, may produce ancestral reconstructions that are either: (1) very general, meaning that many states are potentially included in the ancestors, or (2) not well supported, meaning that one reconstruction is almost as good as another. Bremer (1995), in a different context, discussed the problems involved in using unordered parsimony algorithms for reconstructing ancestral biogeographic character states, and if they seem particularly troublesome to an investigator, steps can be taken to ameliorate them. Specifically, step matrices could be used [see Maddison and Maddison (1992) and Ree and Donoghue (1998)] which specify *a priori* how transitions occur between different regions. (It is conceivable that with this degree of information about the relationships among geographic areas, maximum-likelihood approaches to biogeographic character optimization also might work.) This approach could be used to constrain reconstructions of ancestral states if one were sure that biogeographic transitions from one region to another always involved passing through another region and were much more likely in any case. For example, it might not be a bad assumption, if the areas of interest were Australia, New Zealand, and North America, to treat transitions between Australia and New Zealand as much more probable than transitions between either of those regions and North America. Further, it might also be reasonable to specify that the path of biogeographic movement between Australia and North America always included New Zealand as an intermediate state.

8.5. Areas as Binary Characters

The second major approach to studying the ancestral biogeographic condition of individual clades was developed by Bremer (1992). Here, the aims were different, because determination of the biogeographic states at each of the ancestral nodes was no longer at issue. Rather, it was the biogeographic state of the last common ancestor of the entire clade, the ancestral area, that was being inferred. Further, this approach is somewhat different from the first because geographic occurrence is no longer a single character with several states. Instead, each region or area can be thought of as a binary character, which was either present in the ancestral area or absent from it. Such characters were treated as being irreversible. Bremer (1992) reasoned that for a cladogram of a group one could look at each area or character separately, optimize that character to a tree using Camin-Sokal parsimony, and then see which areas were most parsimoniously explained as being part of the original, ancestral area, or rather as areas that were subsequently entered into. He suggested that the areas on the cladogram that required the fewest independent losses relative to gains would be the ones most likely to be part of the original, ancestral area of the group. This method is also described in Morrone and Carpenter (1994).

Bremer's (1992) basic approach was endorsed by Ronquist (1994, 1995). However, the latter challenged the notion that areas should be treated as irreversible characters that are mapped to a tree using Camin–Sokal parsimony because this assumption is only valid if dispersal is irreversible and a region cannot be subsequently reinvaded. He suggested that there is no reason why we should assume that an area can be invaded only once and no biogeographic processes that would always conspire to produce such a result. Thus, the choice of Camin–Sokal parsimony to optimize individual areas to a cladogram one at a time was flawed. Instead, Ronquist (1994, 1995) argued that allowing dispersals to occur as unordered, reversible events was probably more realistic; therefore, Fitch parsimony should be used to optimize characters (areas) to trees. Then, in his approach, as in Bremer's (1992), those areas that required the fewest independent losses relative to gains on the cladogram would be the most likely to be part of the original, ancestral area of the group.

Bremer (1995) subsequently suggested that there were problems with an approach based on Fitch parsimony because when it is used the difference between accepting one area and another as the ancestral one of a group is generally very small, and this criticism is valid. The question becomes whether one prefers to assume that dispersal is irreversible [which Bremer's (1992) method necessitates] or to accept the fact that when using Fitch parsimony [as in Ronquist's (1994, 1995) approach, which is more agnostic as to the potential for dispersal between areas] it may be hard to differentiate between one area and another as the ancestral one.

8.6. Dispersal–Vicariance Analysis

Ronquist (1997) presented a significant analytical advance in the reconstruction of historical biogeographic patterns in individual groups, in which ancestral distributions are reconstructed with the assumption that speciation occurs by vicariance. In order to explain distributions that do not exactly match a strict vicariance pattern, episodes of dispersal and extinction are invoked. Each of these episodes involves a cost, and the reconstruction that minimizes the cost, while maximizing vicariance, is the one preferred.

One problem with the Ronquist (1997) method is that it always assumes that speciation is due to vicariance, and then invokes a number of dispersal and extinction events to come up with episodes of vicariance. (Note this problem is not unique to this method.) Although vicariant speciation might be expected to play an important role in influencing biogeographic patterns, it would seem more reasonable to have a method of biogeographic analysis that begins with a discovery procedure, to determine to what extent vicariance, dispersal, or extinction occurred during the evolution of a clade. This would be instead of assuming that vicariance is the primary factor governing biogeographic patterns. The role of geodispersal as an important biogeographic process provides a cautionary element to the assumptions made about vicariance.

In a way, although Ronquist's (1997) approach is an analytical advance on his previous work (Ronquist, 1994, 1995), it is somewhat of a retreat from his earlier, more ecumenical posture. His original approach to analyzing the historical biogeography of individual groups treated vicariance and dispersal *a priori* as equally likely, and then sought to determine when in a clade's history a region became occupied. Further, in all three of his papers, he emphasized that dispersal is a frequently occurring phenomenon that must be considered in biogeographic studies. However, the analytical approach of Ronquist (1997) placed primary emphasis on vicariance, while episodes of dispersal and extinction were viewed as secondary factors. Thus, the use of the term dispersal in the couplet dispersal–vicariance analysis, which he coined to describe his method, is a bit of a misnomer. Episodes of dispersal and extinction can still be recovered, but conclusions about how they influence biogeographic patterns in groups would be very different if they were given the same weight as episodes of vicariance. Because his method downplays range expansion it runs the risk of missing episodes of range expansion or geodispersal that are potentially congruent. If, as argued earlier, such episodes do play an important role in influencing historical biogeographic patterns, then Ronquist's (1997) approach may not be complete, or at least it should be utilized with some caution.

8.7. Conclusions

It is clear that there are many quantitative ways to reconstruct biogeographic patterns in individual clades. Repeatability and quantification are important elements of the scientific process, and the fact that each of these is now implicit in biogeography greatly expands the contribution this discipline can make to evolutionary biology and geology. In the next chapter, there will be more discussions of analytical approaches. There, one of the crucial issues in historical biogeography will be considered: how can we analyze biogeographic information from several clades to make more general statements about how geological events have influenced the evolution of biotas.

Chapter 9

The Search for Congruence: Analyzing Biogeographic Patterns in Several Clades

9.1. Introduction

In Chapter 8, I emphasized the historical biogeography of individual clades. Although determining where a single group originated can often be quite interesting, it is not generally the prime thrust of research in historical biogeography, where a research program involves the study of the role of geological and climatic changes in evolution. That is, how do extrinsic factors influence the history of life. As discussed previously, the extent to which we believe evolutionary change is governed by external as opposed to internal factors very much influences our outlook on the evolutionary process in general and on the history of life in particular. Obviously, if external forces drive evolution, then contingency plays an important role in the history of life because that history will be influenced by the history of the Earth. With such a perspective, we might predict that without geological or climatic changes, little evolution will transpire. Biogeographic analyses that integrate patterns from several groups are crucial if we want to know the relationship between the history of the Earth and the history of life.

Biogeographic patterns from a single group cannot be used by themselves to demonstrate the role Earth history events play in motivating evolution, because they could be the result of factors intrinsic to those organisms— ecological requirements or competitive principles or other intrinsic factors that are not immediately apparent. In short, uncovering a biogeographic pattern in a single clade that is compatible with vicariance might suggest that external forces drove the evolution and geographic differentiation of that clade, but it is not enough to be certain. On the other hand, demonstrating the same or similar biogeographic patterns in several groups that occur in the same region provides much stronger support for the notion that this entire biota was shaped by similar Earth history events.

Usually, the aim in a biogeographic study is not just to document that Earth history has governed evolution, but also to determine the sequence of those Earth history events. Again, when several different clades suggest that the same sequence of events influenced evolution, we can be more confident that this was in fact the way it happened. When it comes right down to it, it is a statement about probabilities, although the probabilities are very poorly constrained. They are poorly constrained because there is not yet a quantitative cutoff point beyond which it can be stated with certainty that enough different groups show the same pattern to justify a solid determination that Earth history events prevailed in shaping their evolution. Realistically, it is hard to imagine how such a cutoff could be devised. The basic position advocated here is that the more groups that can be brought into the study of a biogeographic problem the better.

Thus it is clear why one might want to incorporate several groups into a biogeographic study. As stated previously, the basic goal of any historical biogeographic study is documenting the pattern shown in Fig. 2: that different groups showed the same pattern of changes in their geographic distribution as they evolved. However, in the real world, the situation is generally much more complicated, and we need complex analytical techniques to try to tease apart this biogeographic signal from any extraneous noise that may be obscuring it.

9.2. Potential Sources of Noise in Paleobiogeographic Studies

What are some of the types of noise that might obfuscate this historical biogeographic pattern? These have been discussed extensively by many authors including Rosen (1978, 1979), Nelson and Platnick (1981), Platnick and Nelson (1978), Wiley and Mayden (1985), Brooks (1985, 1988, 1990), Wiley (1988*a,b*), Morrone and Crisci (1995), Page (1990), and Brooks and McLennan (1991). From the perspective of an historical biogeographer, noise is any process that disturbs the obdurate biogeographic similarity that should be apparent between different groups in a perfect world. Basically four categories of noise have been invoked, though not all authors recognise all four categories.

All of the authors listed above have suggested that dispersal is something that will obscure a biogeographic signal. This is true in the case of traditional dispersal, which is produced by events relating to the unique ecology of individual clades or individual species within those clades. Such unique ecologies will not be manifested across several different groups, each of which has its own unique ecology and physiology. Instead, they will cause one group to move from one region to another while another group does not, or perhaps moves between different regions. Although this type of traditional dispersal will clearly impart noise to an historical biogeographic signal, I have argued extensively that there can be types of range expansion, which I call geodispersal, that are congruent, replicated across several clades, and related to geological or climatic events. Geodispersal is not a process that overrides an historical biogeographic signal. It is rather part of that signal, and as it is an important historical biogeographic process, any method of biogeographic analysis that does not or cannot take it into account will be inadequate and will probably produce partially spurious results.

Another type of process that can lead to noise in historical biogeographic studies is sympatric speciation, which is in a sense analogous to traditional dispersal, and involves speciation related to unique ecological partitioning within an ancestral species' range. It is difficult except in the most contrived cases to imagine how independent groups of organisms could be expected to show congruent, similar patterns of sympatric speciation in the same regions. Instead, sympatric speciation is related to unique ecological characteristics, and if this was the only way that evolutionary divergence occurs (which is manifestly not the case), it should produce evolutionary patterns very different from those of Earth history events that shape entire biotas.

A third category of noise that various authors have claimed plagues biogeographic studies is generated when a geological or climatic event causes emergence of barriers that lead to evolutionary divergence and speciation in some groups in a region but not in others. Referred to as failure to speciate, it leaves some taxa with wide geographic ranges that circumscribe several areas of endemism, and other species restricted to narrow regions because they have speciated in response to the presence of the barriers. It turns out that widespread taxa and the associated problem of failure to speciate are problematic for some but not all historical biogeographers. As will become apparent, it is most problematic for those historical biogeographers who use components analysis (which will be discussed shortly) to evaluate biogeographic patterns. Those who utilize Brooks parsimony analysis (BPA) (also to be discussed shortly), or approaches based on it, do not view the failure to speciate as producing noise.

Actually, the problem that widespread species cause for components analysis is primarily an analytical shortcoming and symptomatic of a problem with the method. It is hard to understand why the failure to speciate in response to a barrier should be a source of noise. As long as other groups speciated in response to the emergence of the barrier, the relative timing of the event can still be reconstructed. Rather, when such a failure occurs, it should be viewed as if what is going on in this clade is uninformative regarding biogeographic patterns in other clades. It certainly does not cloud the signal, and failure to speciate should not be viewed as a source of noise in historical biogeographic studies. The role of widespread taxa in biogeographic studies will be discussed more fully below in the discussion about BPA.

The final source of noise in biogeographic studies is extinction, which is potentially a serious problem for biogeographic studies that concentrate solely on the extant biota. If many taxa have gone extinct in the various clades being studied, so that what is left in the modern world for analysis is a depauperate biota, we will have an incomplete or even a spurious picture of biogeographic differentiation.

We can, for instance, imagine the following scenario: One clade actually consisted of four species distributed in four areas A, B, C, and D (Fig. 16). However, the species in C went extinct and cannot be detected by a biogeographer who only studies extant organisms. Instead, a sampling of

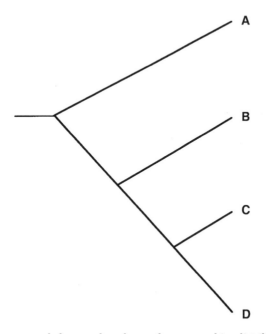

FIGURE 16. An area cladogram based on the geographic distributions and evolutionary relationships of four hypothetical species.

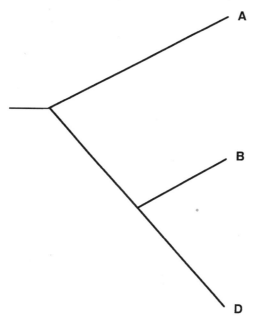

FIGURE 17. Area cladogram that would be retrieved if the species from area C from Fig. 16 is extinct and could not be sampled in a biogeographic study considering extinct taxa.

extant taxa from this clade would produce the area cladogram shown in Fig. 17, which implies that the species in area B is sister to the species in area D. We then imagine that another clade was under study by the same biogeographer and again it consisted of four species, and further that it had the same pattern of area relationship shown in Fig. 16. However, in this second group the species distributed in area D had gone extinct such that it could not be sampled. The apparent area cladogram for this group is shown in Fig. 18. Although in reality both clades diversified across geographic space in concert, when they are converted to area cladograms they incorrectly imply different patterns of biogeographic relationship among regions because there have been episodes of extinction in each clade. This is a simple, contrived scenario, but actually in more complex clades with more complex biogeographic patterns the problem of extinction becomes even more profound.

9.3. Extinct Taxa and the Difference between Biogeography and Paleobiogeography

It is perhaps worthwhile to recognize that every biogeographic study that focuses exclusively on the modern biota is going to come up against the problem of extinct taxa. Any extant clade with a moderately long history is going to have left many extinct taxa behind. With even small or moderate

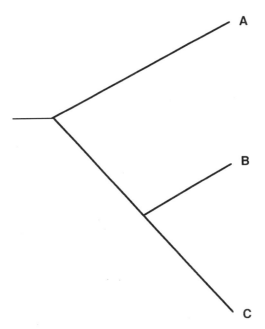

FIGURE 18. An area cladogram produced from another clade considered in the same biogeographic study of extant taxa described in the text and in Figs. 16 and 17. If this clade actually had four species, and the area cladogram matched that shown in Fig. 16, but the species in area D was extinct and could not be studied, then this cladogram and that from the group of Fig. 17 should be congruent, but they are not because of extinction.

degrees of extinction, the pattern of biogeographic area relationship implied by a single clade can change greatly. Combined across several clades considered in a study, extinction can wreak havoc with the output of a biogeographic analysis. In short, potentially any biogeographic study conducted solely on the extant biota can be fundamentally flawed. Those studies that concentrate on groups that have been around for a greater length of time are more likely to be flawed. This simple principle is the must cogent possible argument for paleobiogeography. Without considering the denizens of the fossil record, extinct taxa, biogeographic studies are at risk of serious error.

Paleobiogeographic studies that concentrate largely or exclusively on fossil taxa do not encounter the problem of extinct taxa. In fact, usually all the taxa are extinct! In the fossil record, we are not confronted with a single time slice after long ages of divergence as we are with the extant biota. Instead, clades can be studied as they unfold through time, and representatives from different time slices, still part of the same clade, can be incorporated in the same study. Thus, although the fossil record has at times been denigrated as a source of biogeographical insights (e.g., Nelson and Platnick, 1981; Brundin, 1988), paleobiogeographic studies may have numerous advantages over biogeographic studies that focus exclusively on modern organisms. Now, as argued since Darwin (1859), there are various difficulties associated with the study of evolution in the fossil record. However, since then and particularly since the publication of Eldredge and Gould's (1972) work on punctuated equilibria, numerous studies have documented the relative fidelity of the fossil record, and these were described in Chapter 3. Although caution must be used when interpreting it, the fossil record is a rich source of data that cannot be discarded.

From a paleobiogeographical perspective, the failure of a species to be preserved in the fossil record would be analogous to extinction in the modern biota. If species in various clades are missing from the fossil record, it is conceivable that spurious biogeographic patterns might result, just as can happen if we do not concern ourselves with extinction. Then, just as it is crucial in biogeographic studies of modern organisms to assume that extinction is minimal, in paleobiogeographic studies it must be assumed that the fossil record is fairly faithful, with few missing taxa, and that the probabilities of preservation of all species must be roughly equal. Clearly, the assumption about lack of extinction in modern biogeographic studies is violated in nearly every case. What about the validity of the assumptions needed for paleobiogeographic studies? Those discussed in Chapter 3 provide ample evidence that on the whole the fossil record is fairly faithful, supporting the validity of a research program in paleobiogeography. However, it would be worthwhile to model how changing the preservation probabilities of species in the fossil record would influence paleobiogeographic patterns.

As extinction can potentially influence biogeographic patterns artificially, it is a topic worthy of discussion. Among the authors who have considered the relationship between biogeographic patterns and extinction in greater detail are Brooks and McLennan (1991) and Van Oosterzee (1997). Brooks and McLennan (1991) recognized that the process of vicariant speciation was logically conjoined to extinction: As geographic barriers divide a single large region into two smaller regions, and thus a single large population into two separate populations, which subsequently diverge and speciate, the ancestral lineage will diverge into two new species and thus cease to exist, i.e., extinction is implicit in vicariant differentiation. This point was reiterated by Nixon and Wheeler (1992), who tried to show that generally any mode of speciation must be tied to extinction. Thus, any time an hypothesis of vicariant speciation is invoked within a biogeographic framework there must be some lineage extinction events.

Van Oosterzee (1997) presented a compelling argument concerning the role that extinction plays in generating biogeographic patterns in reference to Wallace's Line. This line, which lies between the islands of Bali and Lombok in the Malay Archipelago, has often been treated as the boundary between the Asian and Australian biogeographic regions. Because Wallace's Line is one of the best known large-scale biogeographic phenomena, the role that extinction may have played in generating this pattern merits further discussion.

Geographically, Bali is part of the Asian or Indo-Chinese plate, but Lombok lies in a geologically unstable region between Asia and Australia (de Boer and Duffels, 1996). Van Oosterzee (1997) suggested that the faunal disjunction between Bali and Lombok, as well as other biogeographic patterns in the Malay Archipelago, were not caused by the types of geological processes and sea-level changes that I have been discussed thus far. Instead, she suggested that these patterns may be attributable to the many active volcanoes in the region. (Such activity can be attributed to tectonic processes, but its influence on biogeographic patterns is very different from those of rifting and continental collision).

For example, the infamous volcano Krakatau is in the Malay Archipelago, and the island of Bali was volcanic. Van Oosterzee (1997) suggested that cataclysmic volcanoes extirpated any faunas that were initially present in the vicinity of what is now Wallace's Line. These faunas may originally have been intermediate in character between what now appear to be the two very disjunct biogeographic regions of Asia and Australia. She suggested that this extinction, by eliminating any clades shared between the two continents, would artefactually make the distinction between Asian and Australian faunas seem profound, when it was in fact not so originally. Thus, in this case, extinctions would have produced biogeographic differences.

Major events such as large-scale volcanic eruptions would potentially influence many clades at once, which would then behave in a congruent fashion in some sense. However, congruence would emerge not due to vicariant speciation or geodispersal but rather owing to wholesale extinction. The prevalence of this type of historical biogeographic mechanism has not yet been demonstrated conclusively in this case nor has it been considered at all in other cases; however, if a similar role for extinction in biogeographic patterns was found to be present in other cases it would substantially alter the way we have to think about congruence in historical biogeography. Other authors who conducted biogeographic analyses on the biotas of the Malay Archipelago suggested that its biogeographic patterns could be explained by the type of tectonic events that we typically associate with biogeographic change, though the tectonic change in this region is complex (Michaux, 1991, 1996; de Boer and Duffels, 1996).

9.4. Additional Basic Assumptions of any Biogeographic Study

Recognizing that there are potential sources of noise in any biogeographic analysis and that their presence may require certain assumptions, we have to make some other assumptions in order to undertake a biogeographic study. These were discussed explicitly by Wiley and Mayden (1985) and Brooks and McLennan (1991): first, there has to be the ability to estimate phylogenies for the groups of interest, and second, it is assumed that character evolution provides a way of inferring the history of speciation.

9.5. Analytical Approaches to Historical Biogeography

Accepting the fact that there is an analytical approach that can be applied to historical biogeographic studies is part of the willingness to accept that there are actual historical biogeographic patterns. These patterns circumscribe many different phenomena, but two of the most important from the perspective of paleobiogeography are the existence of areas of endemism and the close relationship often found between species separated by a geographic barrier. Just as hierarchies of taxa and the existence of taxa are predicted by evolutionary theory, areas of endemism and vicariant distributions are predicted by biogeographic theory, contra Hovenkamp (1997). Each of these phenomena is associated with allopatric differentiation and speciation, which is an overriding theoretical principle in historical biogeography. On the whole, there seems to be overwhelming support for the notion that actual biogeographic patterns exist in the extant biota and the fossil record, and they have been at the core of natural history observations from at least the time of Buffon. If one accepts that such patterns actually do exist, then it is clear that some analytical approach is necessary, but which one?

The situation is similar to the one encountered in phylogenetic analysis, where, if all characters suggested one pattern of evolutionary relationship, then there really was no problem. The problem arose when groups shared different characters. The same principle is true in biogeography: If all clades unequivocally showed the same biogeographic patterns then there would be no need for analytical methods, but, in fact, some clades suggest a different history than others for the areas they occur in (Lydeard *et al.*, 1995). This history reflects the way those areas are related, either by the formation of barriers, which fragmented and vicariated biotas, or by the removal of barriers, which allowed biotas to merge.

It is clear that when we are confronted with this disagreement or incongruence we must choose some analytical method to resolve it. Analytical approaches are essential in any scientific study, for without them, an element of testability and rigor is lost. These analytical techniques do not in themselves guarantee valid results, but without them biogeographic studies are at risk of deteriorating into a series of just-so stories. One biogeographer can support one pattern on the basis of one or two pieces of data perceived to be important, whereas another can support another pattern based on other data also perceived to be important. Without some analytical arbiter, it is impossible to choose between competing systems, and agreement becomes more a statement of belief and taste than an exercise in testing hypotheses (Henderson and Heron, 1977). Bowler (1996) demonstrated how the inability to test competing hypotheses in evolutionary biology retarded progress in early phylogenetic studies and resulted in rancorous debates that could never be settled. This eventually caused systematic studies to lose prestige in the scientific community. By analogy, the same is true of biogeography.

Fortunately, the use of analytical methods has a long and rich tradition in biogeography. Almost from the inception of the field biogeographers tried to employ quantitative methods. Alexander von Humboldt (see Chapter 4) was probably the first researcher to utilize a rigorous analytical method in biogeography. He proposed a science of botanical arithmetic, where the percentage of different types of species in different regions could be compared. Robert Brown (1814) used a similar type of quantification, where the region with the most genera was likely to be the site where the entire family originated (Browne, 1983). De Candolle père et fils, von Buch, and Watson did similar types of analyses. In undertaking this early quantitative approach to biogeography, these natural historians sought to adapt rigorous analytical techniques to make their discipline more quantitative and thus more acceptable to the broader scientific community. In a sense, they wanted to be statisticians or biological census takers (Browne, 1983).

Today, more than ever, it is important for biogeography to be quantitative and rigorous; otherwise it will not be taken seriously as a science. Many rigorous analytical methods have been developed by ecological biogeographers, and these are discussed more fully in MacArthur and Wilson (1967) and Brown and Lomolino (1998). However, this book is primarily about historical biogeographic patterns, which are more amenable to study in the fossil record. Owing to the differences between the entities of the genealogical and economic hierarchies and the fact that processes at one hierarchical level cannot be extrapolated to higher-level entities or across hierarchies, the rigorous methods of ecological biogeography cannot always be applied in a straightforward way to historical biogeographic studies (see Chapter 2). This just reflects the difference between genealogical and ecological entities.

After the development of the so-called early botanical arithmetic at the start of the 19th century, there were not many analytical advances in biogeographic studies, specifically historical biogeographic studies, for a long time. That is not to say that no good work came out of many of the biogeographic studies carried out during the 19th and 20th centuries. Important work was done, including that described in the historical treatments by Mayr (1976) and Bowler (1996), just to mention a few, and of course there were the biogeographic studies of Wallace and others. However, that work was not always highly quantitative. Typically, quantitative approaches in such studies involved comparisons of shared percentages of taxonomic similarity, usually conducted at the generic level. Those regions that shared high proportions of genera in common were treated as sharing a close geological relationship. These approaches were phenetic in nature, and as such were problematic.

9.6. Phenetic Approaches to Biogeographic Analysis

The next major advance in quantification involved the application of techniques such as principal components analysis, cluster analysis, and nonmetric multidimensional scaling to problems in biogeography. These multivariate statistical methods quantified similarities among different regions in terms of the number of taxa, frequently genera, that they shared. Typically, in such methods information about the distribution of taxa is entered into a data matrix for each of the various regions or areas of endemism. This data matrix is then analyzed using a statistical package. If cluster analysis is used, regions are grouped together onto a tree. When regions are close on a tree, it is inferred that they shared a more recent geological history than other regions more distantly positioned.

When principal components analyses is used, regions are mapped as a scatter of points on principal components axes. Those closer to one another on the axes are then treated as sharing relatively close biogeographic affinity. Examples of studies that used these types of multivariate statistical approaches to biogeography can be found in Whittington and Hughes (1972) and Babcock (1994), who used nonmetric multidimensional scaling; Williams (1973) and Babcock (1994), who used cluster analysis; Jell (1974), who used principal components analysis, and Rowell *et al.* (1973), who used all of these methods. The application of these techniques to problems in biogeography was an important breakthrough because it opened the way for rigorous statistical tests of biogeographic hypotheses. Further, results were presented in a mathematical context, the framework for which was well understood.

However, there are problems with these approaches, which make it clear that we have to be cautious when we apply multivariate statistical techniques to biogeographic data. First of all, many of the similarity coefficients utilized in phenetic studies are highly biased and inaccurate (Henderson and Heron, 1977). Second, there are significant problems that arise because of the phenetic nature of these methods. Briefly, as I discussed in Chapter 8, phenetics uses autapomorphies and plesiomorphies to make statements about the relationships of organisms. These characters do not in fact convey any information about patterns of relationship among organisms, and it has been recommended that phenetic techniques be avoided in phylogenetic studies (Eldredge and Cracraft, 1980; Wiley, 1981; Brooks and McLennan, 1991; Wiley *et al.*, 1991; Smith, 1994). When multivariate statistical techniques are applied to biogeographic studies, biogeographic regions are grouped by patterns of phenetic relatedness. Taxa that were primitively distributed across a region or that occurred uniquely in a single region influence the patterns of relationship between regions. Thus these techniques are problematic for the same reasons that phenetic techniques in systematics are troublesome. They may group things on the basis of spurious criteria that, in fact, contain no information about patterns of relationship.

Another problem with these techniques is that regions are grouped on the basis of similarities in their taxa, but other than that, no evolutionary information constrains biogeographic hypotheses. For example, when several genera occur among different regions, the fact that some of them may be closely related to one another and others more distantly related would not be factored into phenetic studies, although this information can reveal something important about biogeographic patterns. However, it is factored into the phylogenetic approaches to biogeography, which are described more fully below. In spite of these difficulties, it is clear that these techniques represented a real advance, and also constituted an important step in the development of biogeographic analysis.

9.7. Probabilistic Approaches to Biogeographic Analysis

One of the most interesting and novel approaches to biogeographic analysis was developed by Henderson and Heron (1977), who tried to place the analysis of patterns of distribution of organisms in the fossil record into a probabilistic framework. They also emphasized the great potential of paleobiogeography as a discipline, arguing that paleontologists should not just accept existing paleogeographies derived by geophysical techniques but rather actively test them using paleobiogeographic data. Their method was based partly on phenetic techniques since they looked at the number of taxa shared among different regions, but they treated it as a problem in probability. Specifically, they considered the number of ways that a similar number of species can be selected from an underlying uniform population.

Although their emphasis on phenetic techniques suffers from the same criticisms I discussed above, Henderson and Heron's (1977) approach was very creative and insightful. Further, it may have real potential for determining the boundaries of areas of endemism. As I discussed in Chapter 7, this exercise is currently plagued with significant problems and lacks analytical rigor. Extending their method to this problem might make paleobiogeography more quantitative as its strength is that it easily lends itself to a hypothesis-testing framework. Unfortunately, there has as yet been no follow up on this work.

9.8. Analytical Approaches to Biogeography within a Phylogenetic Framework

There is a basic disjunction between the analytical approaches to biogeography just discussed and those that utilize phylogenetic information. The latter not only use data relating to the distribution of taxa, as both the phenetic and probabilistic methods do, but they also incorporate information about how the distribution of taxa changes as the clade they are in evolves. Thus phylogenetic approaches lend themselves to a much more rigorous search for biogeographic congruence.

When phylogenetic approaches are used in biogeographic studies the results are presented as a tree relating different areas. Different regions are grouped into a hierarchy of relatedness, and those areas that are most closely related on such trees are treated as sharing a very recent common geological history. More distantly related areas had a common geological history deeper in the past. In this tree-based approach, the situation is analogous to a tree produced by phylogenetic analysis. However, instead of displaying a shared history of common descent, the shared history is one of geological connectedness.

One of the other features that distinguishes phylogenetic from phenetic approaches is that, for philosophical reasons, separate regions are not grouped on the basis of their possession of unique (autapomorphous) taxa or taxa that are primitively present in regions. Another commonality that some of the phylogenetically based methods share is that they use a parsimony-based algorithm in data analysis.

At present, there are two basic methods of biogeographic analysis that are based on a phylogenetic approach: components analysis and Brooks parsimony analysis (BPA).

9.8.1. Components Analysis

Components analysis was initially developed by Platnick and Nelson (1978) and subsequently extended by them (Nelson and Platnick, 1981), although Rosen (1978, 1979) developed similar methods. Components analysis converts cladograms for different groups of organisms into area cladograms by substituting the area(s) of endemism for the species name and then compares them for similarities and points of departure. These different area cladograms may diverge from one another because not all areas are represented in the different cladograms, as some taxa may be distributed in several areas or because some clades indicate different patterns of area relationship. The method tries to reconcile these divergences using two procedures: first, assumptions about aspects of the data are made for each group; and second, a consensus technique is used to find commonalties among the clades after the assumptions are made. In the first procedure every area cladogram is examined for departures from adherence to strict vicariant differentiation, which would include any widespread taxa or any regions that appear more than once on the tree. Assumptions are then made about those areas that do depart, and the cladograms are altered accordingly. There are two possible assumptions for the data: assumption 1 or assumption 2, described in detail in Nelson and Platnick (1981) and Morrone and Crisci (1995). Basically, they involve generating several new area cladograms for each of the original ones by adding new branches at different nodes. Where and how many of these branches are added differs according to which assumption is made. After all the new area cladograms are generated, there is a search and any tree(s) found to be shared in common among the different area cladograms is taken as the best pattern of area relationship. If several trees are shared in common, a consensus tree is generated, which describes the common elements that the multiple trees share. If no trees are shared in common, a consensus tree (see Wiley *et al.*, 1991) of all of the trees can be generated (Morrone and Crisci, 1995).

There are reasons to believe that there are significant problems with each of the procedures that components analysis employs. First, there are difficulties with the assumptions, which have been convincingly criticized by Wiley (1988*a,b*) and Brooks and McLennan (1991). Basically, assumptions are made about some but not all of the data represented in the area cladograms, but, in principle, there is no reason why only the elements that depart from strict vicariance should be questioned. It would be equally plausible to assume that the data compatible with vicariance are also invalid. This approach violates one of the fundamental philosophical principles underlying the discipline of phylogenetic systematics, which was taken as a logical necessity in order to avoid hypothesizing rampant parallelism when discerning evolutionary relationships. Hennig (1966) suggested that investigators should initially treat any characters shared between taxa as indicating shared common descent, which amounts to initially treating data as real (Wiley, 1988*a,b*). By violating this principle, components analysis is at risk of serious bias. Its use makes it impossible to discern any pattern not compatible with strict vicariance. Thus, geodispersal could never be recovered as an important biogeographic process.

Components analysis also suffers from the use of consensus techniques, which are methods that are applied when there are several equally well-supported trees and we want to see how they agree (Wiley *et al.*, 1991). Consensus techniques produce a single tree by combining information from multiple trees, but if the different trees disagree in many of their elements the consensus tree (depending on which of the many types of consensus approach is utilized) will have little or no resolution among the groups. It has been repeatedly demonstrated (e.g., Miyamoto, 1985; Barrett *et al.*, 1991, 1993) that when consensus techniques are used in phylogenetic analysis, the best-supported, most parsimonious pattern is often not recovered. These problems make it clear that consensus techniques should be used with caution in biogeographic studies. When components analysis implements a consensus technique to square several competing biogeographic hypotheses it may have the same drawbacks as such techniques suffer in phylogenetic analysis (Kluge, 1988; Wiley, 1988*a,b*).

There is still another problem implicit in the application of consensus techniques to biogeographic studies. Essentially, in components analysis each area cladogram from each group for which a phylogeny exists is treated separately. These cladograms are then augmented by applying certain assumptions. Consensus techniques may be necessary to synthesize the results from the cladograms for each of the different groups, which leads to each group being given equal weight in the biogeographic analysis. The area cladograms for many groups might be fairly similar in their overall pattern, but a single group that is different can alter the entire biogeographic pattern by causing loss of resolution in the consensus tree. It would seem to be better to let different groups contribute different amounts of biogeographic signal, rather than to treat each group as of equal weight *a priori*. If an overarching biogeographic pattern is well supported in several but not all the groups application of any method that fails to adequately resolve biogeographic patterns seems problematic. That is not to say that components analysis can never be applied in scientifically valid or intriguing ways. Page and Lydeard (1994), Lydeard *et al.* (1995), and de Boer and Duffels (1996) have illustrated that it can be used successfully in certain cases. Still, apart from the fact that it is based on questionable assumptions, it will be difficult or impossible to use components analysis to study important biogeographic processes such as geodispersal, so it cannot be recommended as a general discovery procedure in biogeographic analysis.

9.8.2. Brooks Parsimony Analysis (BPA)

The Brooks parsimony analysis (BPA) technique was first developed for use in biogeographic and coevolutionary studies by Brooks (Brooks *et al.*, 1981; Brooks 1985), and later extended in several other publications (Brooks, 1988, 1990; Wiley, 1988*a,b*; Brooks and McLennan, 1991; Wiley *et al.*, 1991; Lieberman and Eldredge, 1996; Lieberman, 1997). Similar to components analysis, BPA uses information about the geographic distribution of taxa, as well as their phylogenetic relationships, to infer biogeographic patterns, and the first step is to obtain cladograms, preferably from several different groups of organisms. The geographic occurrence of each taxon is then substituted for the taxon names on each of these cladograms, but after that the two methods diverge significantly.

In BPA, the next step is an optimization procedure to determine the biogeographic state of each of the ancestral nodes for the cladogram akin to what was discussed in Chapter 8 (Fig. 19). Then, a data matrix is generated, its rows being the names of each of the geographic areas of interest. [The way the data matrix is generated has changed as BPA has evolved as a technique, but the basic principles, discussed in Wiley *et al.* (1991), still apply; a modification of this method recommended herein is presented more fully below.]

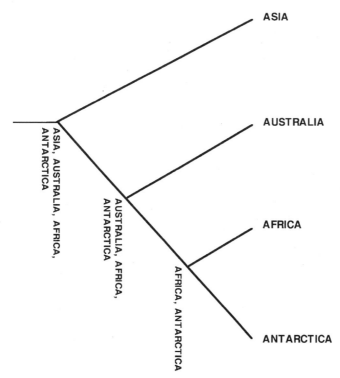

FIGURE 19. An area cladogram for four hypothetical taxa with the biogeographic states of nodes optimized using the exclusive OR-ing method employed by standard BPA.

Table 2. Traditional BPA Coding for Area Cladogram
shown in Fig. 19.[a]

	1	2	3	4	5	6	7
Ancestor	0	0	0	0	0	0	0
Asia	1	1	0	0	0	0	0
Australia	1	0	1	1	0	0	0
Africa	1	0	1	0	1	1	0
Antarctica	1	0	1	0	1	0	1

[a]"0" represents absence from a region; "1" represents presence in a region. Ancestor refers to the ancestral biogeographic condition for the clades considered. The rows represent the areas being analyzed, and the columns are biogeographic data, where characters 1, 3, and 5 are the three nodes, ascending up the tree, and characters 2, 4, 6, and 7 are the terminal taxa.

BPA codes information about the distribution of each of the ancestral nodes and each of the terminal taxa into what is called a BPA matrix. Every node and every terminal taxon is a column of the matrix. If a terminal taxon or a node is present or absent in a geographic region, a "1" or a "0", respectively, is entered in the appropriate row, as illustrated in Fig. 19 and Table 2. Presence in two areas is taken as evidence that these regions shared a common geological history and were once joined together. Thus, unlike in components analysis, the information from widespread taxa, be they ancestral nodes or terminal taxa, is utilized. Treating a taxon's presence in more than one region as evidence for the relatedness of these regions has been referred to as assumption 0 by Zandee and Roos (1987) to distinguish it from assumptions 1 and 2 of components analysis. They called it assumption 0 because they saw it as equivalent to accepting the biogeographic data at face value, unlike assumptions 1 and 2, and took it to be fundamentally more valid. If one is not sure if a terminal taxon or an ancestral node was present or absent from a region, a "?" can be placed in the data matrix instead of a "0" or a "1."

After the data matrix is generated, it is analyzed using the computer algorithm PAUP 4.0 (Swofford, 1998), which is the same algorithm utilized in a phylogenetic study of organisms involving either DNA sequences or morphological characters. In fact, its application in biogeographic studies is exactly the same as in phylogenetic studies. To implement this program, another additional row has to be added to the data matrix, which is typically given the name "Ancestor." From this region, all the terminals and ancestors are primitively absent, so it is a row of all "0's." This is a simple procedure, which is carried out so that PAUP knows that "0" represents absent from a region (the primitive state) and "1" represents present (the derived state).

9.8.3. Synthesizing Results in BPA and Why a Parsimony Algorithm Is Used

The by-product of BPA will be a cladogram of area relationships, which is exactly analogous to one relating different taxa. In the latter, the information concerns how recently the taxa last shared a common ancestor, while in the former it relates to the common history that the areas share. The more closely related two areas are seen to be, the more recently they shared a common geological history. Thus evolutionary trees and area cladograms both specify common history; in the former, it is shared common ancestors; in the latter, shared geological history, which might relate to the time that these now separate regions were joined.

The parsimony algorithm is used to recover the congruent signal in the data and reveal the pattern of area relationships; the philosophy behind this approach is worth considering. Clearly each clade, and the species within it, has its own unique ecological characteristics that partly governs how a group evolves and its geographic range changes. Unless different clades have precisely the same ecological characteristics, we would expect a different geographic distribution in each clade. If individual ecologies were what most influenced biogeographic patterns, we would expect each clade to show an individualistic biogeographic response and biogeographic patterns to differ among clades.

How does BPA uncover congruent or incongruent biogeographic patterns? We can imagine an example where two different clades show incongruent biogeographic patterns because the changes in geographic distribution in each were most influenced by the individual ecological characteristics of their component taxa rather than by Earth history events. For simplicity, we assume that there are two characters in the BPA matrix, one for each clade. The first character from the first clade involves an ancestral node that was found in only two regions, A and B, which would be captured by coding a "1" in regions A and B in one column. To PAUP this would be a synapomorphy uniting these two regions, which implies that they shared a common geological history when they were joined together. Now, we imagine that the second character represents an ancestral node of the second clade that was found in regions A and C but not in B. This node would be coded into the matrix as "1" in A and C, which, again, PAUP would interpret as a potential synapomorphy.

With these two characters, there are conflicting patterns of synapomorphies, one suggesting that A and B share a unique common geological history, and the other that B and C share such a history. If each character has equal weight, there is no way of choosing between these conflicting characters and the relationship among areas A, B, and C cannot be resolved. In such a situation, where individualistic ecological factors of clades dominate the speciation process, this is what we would expect to find. In a real-world study where these kinds of factors prevail, we would also expect that there would be little or no resolution in a tree of area relationship. Even if the area tree is resolved, we would expect the area cladogram to be poorly supported by various relevant measures such as bootstrap values, jackknife values, permutation tail probability tests (Faith, 1991; Faith and Trueman, 1996), tree-length frequency skewness distributions (Hillis, 1991), and Bremer branch-support values (Bremer, 1994).

When Earth history events control biogeographic patterns we would predict a very different scenario, in which many groups would show the same pattern of differentiation across geographic space. However, it would be naive to believe that every group would show precisely the same pattern. There would be differences from group to group. Parsimony, when applied to a BPA matrix, will uncover the most frequently replicated pattern. An example using the three regions from the previous discussion, A, B, and C will help illustrate this. Now we imagine that we are considering three clades. In two of them, there is an ancestral node distributed in areas A and B. In BPA this will be captured by coding two characters as present in A and B and absent from C, leading to two synapomorphies indicating that these regions have a unique shared geological history.

In the other clade, we imagine that there is an ancestral node distributed in areas A and C, which will be captured in BPA by coding a single character as present in A and C and absent from B. This is a single synapomorphy indicating that these regions have a unique shared geological history. When it comes time to analyze this matrix using the parsimony algorithm PAUP the grouping of areas A and B will be more strongly supported than the grouping of areas A and C because there are two synapomorphies supporting the former relationship and only one supporting the latter. If Earth history processes are the dominant influence controlling biogeographic patterns, then the resulting area cladogram should be well resolved, as is the case here. Further, the various measures of support for this tree, noted above, should now indicate that it is well supported.

In the end, parsimony is a way of seeing which regions are most consistently shared among ancestral nodes and terminal taxa. There will be some disagreement from clade to clade or even within the same clade, but parsimony can determine the overriding congruent pattern. The principle behind its use is an attempt to uncover the shared signal. The incongruencies from clade to clade are certainly important, but they do not provide information about how Earth history processes influenced the biogeographic patterns in these regions.

9.8.4. Modifications to BPA in Coding

9.8.4.1. Optimizing Nodes

We have proposed two modifications to the method I have just described of coding a BPA matrix, and one more fundamental modification of BPA that relates to the nature of the analysis (Lieberman and Eldredge, 1996), all of which are discussed more fully below. These modifications which involve changing the way the ancestral nodes are optimized, make it possible to capture additional information about biogeographic patterns. In classic BPA [see, e.g., Brooks and McLennan (1991) and Wiley *et al.* (1991)] the distributions of the ancestral nodes are determined by an optimization procedure called inclusive OR-ing, which works by proceeding from the tips of the tree to the root. The geographic states of ancestral nodes are the summation of the states of all of their descendants (see Fig. 19). This type of optimization sometimes leads to idiosyncratic results; it also guarantees that as one proceeds up any cladogram from the root to the tips the geographic range of the ancestral nodes will slowly and regularly contract. This is equivalent to a straightforward pattern of vicariance, because every step up the tree from ancestral to descendant node, which can be thought of as a speciation event, involves a contraction in range. (It should be recalled that in vicariant differentiation, speciation involves dividing a larger range of an ancestral species up into smaller ranges of descendant species that will, in combination, be equal to the range of the ancestor.) Thus, when inclusive OR-ing is used, vicariance is the only signal that can be recovered from the data. Clearly vicariance is an important biogeographic process, but as I have argued throughout this book, it is not the only biogeographic process worthy of study.

We (Lieberman and Eldredge, 1996) proposed a coding procedure that we believe makes fewer *a priori* assumptions about biogeographic patterns and also obviates the idiosyncratic results that are discussed below. This method utilizes the Fitch (1971) unordered algorithm of character change described in Chapter 8. Liebherr (1988), Enghoff (1995), and Ronquist (1995) have provided a general endorsement of the use of Fitch parsimony in historical biogeo-graphic studies. By assuming that biogeographic state changes are equally likely to occur among all regions, the Fitch algorithm does not bias the results to a vicariance pattern *a priori*, though it allows such a pattern to be retrieved if it is compatible with the data. Thus, for a taxon cladogram, the optimization would look like Fig. 20. In this case, an ancestral biogeographic state was mapped to the putative root of the clade of interest prior to performing the Fitch optimization. This has heuristic value [see the discussion in Ronquist (1995)] if we know something about the phylogenetic origins of this single clade at a broader contextual level, which might prove informative regarding biogeographic patterns.

9.8.4.2. Translating Area Cladograms to a BPA Data Matrix: Coding Procedures and Capturing Both Vicariance and Geodispersal

Our second modification to BPA (Lieberman and Eldredge, 1996) related to how the information from the area cladograms was coded into the data matrix, and expanded the types of biogeographic processes BPA might capture by adding a procedure for capturing both vicariance and geodispersal. The starting point for this modification was based on the fact that an area cladogram with optimized nodes encodes several different types of informa-tion. First, there are the geographic distributions of the ancestral nodes and the terminal taxa. However, when we look at the optimized area cladogram we can see how the geographic distributions change as the group evolves. Specifically, between ancestral and descendant nodes and between ancestral nodes and descendant terminal taxa the geographic range changes frequently. These data bear on what happened to geographic range as an ancestor differentiated into its descendants. These types of patterns are fundamental to historical

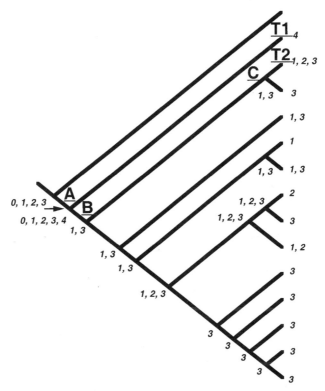

FIGURE 20. Area cladogram for the trilobite genus *Basidechenella* with biogeographic states substituted for terminal taxa and mapped onto the ancestral nodes using the technique described in the text: 0 = Canadian Arctic, 1 = Appalachian Basin in eastern North America (ENA), 2 = Illinois Basin in ENA, 3 = Michigan Basin in ENA, and 4 = Armorica. Additional symbols are explained in the text. Arrow denotes the ancestral biogeographic state of the group. From Lieberman and Eldredge (1996), used by permission.

biogeography, as they tell us about geographic range in an evolutionary context, either directly or secondarily related to the speciation process, depending on the taxonomic category of the taxa in the area cladograms. A method of coding BPA matrices that can capture this type of information can enhance the ability of BPA to recover biogeographic patterns. One of our modifications to traditional BPA (Lieberman and Eldredge, 1996; Lieberman, 1997) was an attempt to capture this information.

Further, as discussed earlier, there are really two types of congruent Earth history events that can influence biogeographic patterns. One is the rise of geographic barriers owing to geological or climatic events, which leads to congruent patterns of allopatric speciation in several different clades. These equate to vicariance. There are also times when geographic barriers fall owing to geological or climatic events, which leads to congruent patterns of range expansion in several different clades. These are equivalent to geodispersal. Both are fundamental events in the history of biotas. A method of biogeographic analysis that can capture these types of events should have strengths over approaches that only consider vicariance. We argued (Lieberman and Eldredge, 1996) that to capture congruent episodes of vicariance and geodispersal any biogeographic analysis had to be divided into two separate analyses—one that looked for congruent episodes of vicariance and one that looked for congruent episodes of geodispersal—and each analysis would have its own BPA matrix. If these types of biogeographic information are not considered separately then the biogeographic signal implicit in the data will be lost: As the patterns of change in geographic distribution produced by vicariance are different from those produced by geodispersal, these patterns will be conflicting and self-effacing because the pattern of contraction in range associated with vicariant differentiation will be overshadowed by a subsequent episode of geodispersal.

To capture the information relating to how geographic distributions changed as the group evolved, as well as to search for congruent episodes of both vicariance and geodispersal, we (Lieberman and Eldredge, 1996) proposed some simple criteria based on the fact that between adjacent ancestral and descendant nodes and between ancestral nodes and their terminal taxa in any area cladogram there are basically three things that can happen to the geographic range: it can contract, expand, or remain constant. The first is compatible with vicariant differentiation and the second with geodispersal, while the third option implies neither vicariance nor geodispersal. One of the modifications to BPA that we proposed (Lieberman and Eldredge, 1996) was that in the biogeographic analysis that attempts to recover congruent episodes of vicariance only contraction and constancy of range should be considered and coded into a BPA matrix. This will produce what is called a vicariance matrix because it is designed to retrieve congruent episodes of vicariance. We also suggested that in the biogeographic analysis that attempts to recover congruent patterns of geodispersal, only expansion and constancy of range should be considered and coded into a BPA-type matrix; this will produce what is called a geodispersal matrix because it is designed to retrieve congruent episodes of geodispersal.

9.8.4.3. Coding the Vicariance Matrix

To illustrate how our method (Lieberman and Eldredge's, 1996) for generating a vicariance matrix works, I will translate the information about how geographic distribution changed during the evolution of this clade, the information from the area cladogram shown in Fig. 20 above, into a BPA matrix. For simplicity, I will start with the basic setup of a BPA matrix. Each node and each terminal taxon represents a column of the data matrix, and the different areas are the rows. This matrix has the same format as a character matrix in phylogenetic analysis, except that this time we are not evaluating the evolutionary relationships of taxa but rather the geological relationships of areas, and these are the rows of the matrix. In both types of analyses there is a row, as in phylogenetic analysis, occupied by the outgroup. The columns are not morphological or molecular characters being used to evaluate the evolutionary relationships of the taxa, as was the case in phylogenetic analysis, but rather the ancestral nodes and the terminal taxa of the area cladogram. These provide information about how geographic distributions have changed as taxa have evolved, which are the fundamental data of historical biogeography.

For coding the vicariance matrix, we begin with the coding for the basal-most node of the cladogram, the ancestral biogeographic state of the group, which is marked with an arrow in Fig. 20. This will be the first column or first character of the data matrix shown in Table 3. This ancestor is present in areas 0, 1, 2, and 3 and absent from area 4. There has been some discussion (e.g., Wiley, 1988*a,b*; Brooks and McLennan, 1991; Wiley *et al.*, 1991) about whether or not failure to recover a taxon or its ancestor in a region should be recorded in the matrix as true absence, "0," or just the failure to recover the taxon, "?." The latter choice is certainly the more conservative one, but when there are many "?'s" in a data matrix, it can lead to spurious results (Brooks and McLennan, 1991; Waggoner, 1996; Lieberman, 1998). For this reason, the use of "?'s" in BPA is suggested only if there is a fair degree of uncertainty about a taxon's presence in a region.

In terms of paleobiogeographic studies, which emphasize the fossil record, the use of "?'s" might possibly be recommended if this region were poorly sampled or had a very fragmentary fossil record. As a general rule, absence should be taken as true absence and coded with a "0." This practice seems justified to avoid spurious patterns of relationship emerging owing to the way computer algorithms deal with the "?" character coding. An approach that treats absence as true absence is justified in regions that have a good fossil record, measured in terms of overall stratigraphic completeness, and have been adequately sampled by paleontologists. All codings of characters in this exercise treat absence as true absence, and they will also include the state of the outgroup, which is "0." The first node is coded as 0, 1, 1, 1, 1, 0 in the first column of Table 3.

Table 3. Modified BPA Coding of the Vicariance Matrix for the Basal-Most Node, Nodes A–C, and Terminal Taxa T1 and T2 from the Area Cladogram shown in Fig. 20[a]

	1	2	3	4	5	6
Ancestor	0	0	0	0	0	0
Area 0	1	1	1	1	0	0
Area 1	1	1	1	2	1	1
Area 2	1	1	1	1	0	1
Area 3	1	1	1	2	1	1
Area 4	0	1	2	1	0	0

[a]"0" represents absence from a region; "1" and "2" represent presence in a region. Ancestor refers to the ancestral biogeographic condition for the clades considered. Multistate characters are ordered (additive).

   The biogeographic state of the node directly up the tree, node A, is present in areas 0, 1, 2, 3, and 4. This node will be coded as character 2 in Table 3 with states 0, 1, 1, 1, 1, 1. Note that the transition from the ancestral biogeographic state of the group to node A marks an expansion in range. The transition from node A to the terminal taxon (marked T1), present only in region 4, marks an episode of contraction in geographic range in conjunction with diversification. This is equivalent to vicariance, because as the ancestor diverged into one of its descendants, its range transformed from present in regions 0–4 to present only in region 4. Here, the vicariance event separated region 4 from regions 0–3. To capture this information, a special coding is employed in the BPA vicariance matrix: an ordered multistate character. This character, character 3 of the matrix, is coded as present in regions 0–4 but with a special derived presence in region 4. For this column of the matrix, for regions 0–3 a "1" is inserted, and for region 4 a "2" is inserted. Then, character 3 is treated as an ordered character. The use of the ordered character in BPA was suggested, although in a different context, by Mayden (1988).

   As we proceed up the tree, the biogeographic state of the next node, which is node B, is present in regions 1 and 3 and absent from the other regions. Again, this is a contraction in overall range relative to its immediate ancestor, node A, and thus evidence for vicariance. To denote this, this character, character 4, is coded as an ordered multistate character, present in areas 0–4 but with a special derived presence in areas 1 and 3 or 0, 1, 2, 1, 2, 1.

   The transition from node B to node C involves no change in geographic range so the node is simply entered into the matrix as character 5: 0, 0, 1, 0, 1, 0. The transition from node C to its descendant terminal taxon, marked T2, involves an expansion in geographic range. This change does not conform to an overall pattern of vicariance so an ordered multistate character is not used in character 6. Instead, character 6 is coded simply to reflect presence in areas 1, 2, and 3: 0, 0, 1, 1, 1, 0. If we filled in the codings for the rest of the tree, we would be ready to analyze the vicariance matrix with the parsimony algorithm PAUP. Again, just to summarize, the rationale behind the creation of this matrix was based on the principle that repeated patterns of contraction in an area occupied by nodes and terminal taxa provide evidence for vicariance.

9.8.4.4. Coding the Geodispersal Matrix

   For the same area cladogram we can now generate the geodispersal matrix. Coding is very similar to the vicariance matrix except that now transitions between adjacent ancestral and descendant nodes and between ancestral nodes and their descendant terminal taxa that involved change or expansion of a lineage's range are treated as potential evidence for geodispersal and therefore coded as ordered multistate characters. To illustrate the coding of this matrix we refer again to Fig. 20 and to Table 4. The ancestral state is again coded as character 1: 0, 1, 1, 1, 1, 0. The transition from the ancestral biogeographic state of the group to node A involves an episode of range expansion during the clade's biogeographic history with expansion from

Table 4. Modified BPA Coding of the Geo-Dispersal
Matrix for the Basal-Most Node, Nodes A–C, and Terminal
Taxa T1 and T2 from the Area Cladogram shown
in Fig. 20[a]

	1	2	3	4	5	6
Ancestor	0	0	0	0	0	0
Area 0	1	1	0	0	0	0
Area 1	1	1	0	1	1	1
Area 2	1	1	0	0	0	2
Area 3	1	1	0	1	1	1
Area 4	0	2	1	0	0	0

[a]"0" represents absence from a region; "1" and "2" represent presence in a
region. Ancestor refers to the ancestral biogeographic condition for the
clades considered. Multistate characters are ordered (additive).

present in regions 0–3 to present in regions 0–4. To denote the expansion of
the descendant into area 4 this character, character 2, is treated as an ordered
multistate character and coded as present in regions 0–4 but with a special
derived presence in region 4, denoted by "2." Therefore, the coding for
character 2 is: 0, 1, 1, 1, 1, 2.

The transition from node A to terminal taxon T1 marks a contraction in
range and thus a potential episode of vicariance. This transition is not coded as
an ordered character in the geodispersal matrix. Instead, terminal taxon T1 is
coded in as character 3, present in area 4: 0, 0, 0, 0, 0, 1. The transition from
node A to node B also involves a contraction in overall range. Thus, character 4
is coded as present in regions 1 and 3 or: 0, 0, 1, 0, 1, 0. In the transition from
node B to node C the overall range does not change so character 5 is also coded
as: 0, 0, 1, 0, 1, 0. Finally, the transition from node C to terminal taxon T2
involves an expansion of range, as the taxon has moved from present in areas 1
and 3 to present in areas 1, 2, and 3. Therefore, this is denoted as character 6,
which is coded as 0, 0, 1, 2, 1, 0. If we were to fill in the codings for the rest of
the tree, we would be ready to analyze the geodispersal matrix with the
parsimony algorithm PAUP. To summarize, the rationale for creating this
matrix was that repeated patterns of range expansion into the same area
provide evidence for geodispersal (Lieberman and Eldredge, 1996).

9.8.4.5. Implementing the Parsimony Algorithm

It is recommended that whenever possible each data matrix be analyzed
using the exhaustive search option of PAUP (Swofford, 1998) because this is
the most rigorous option available. Above a certain number of areas exhaustive
searches become prohibitive owing to the time they require, but typically this
number exceeds the number of areas considered in biogeographic analyses.

Analysis of each data matrix will produce one or more most-parsimonious
trees. The ensuing area cladograms from the analysis of the vicariance
and geodispersal matrices are the vicariance and the geodispersal trees,

respectively. Each of these trees relates areas and allows for ready interpretation of biogeographic patterns.

9.8.4.6. Interpreting the Vicariance Tree

The vicariance tree provides information about the relative time that barriers emerged between regions, leading to vicariance. Areas that share a recent common ancestor or node on the area cladogram were fairly recently joined as a single area, and the newest geographic barriers to form were those that separated these regions from one another. Areas that are more distantly related on the tree were joined more distantly in the past, the barriers dividing these areas appearing relatively long ago. In this way, area cladogram topology can be used to infer the timing of the separation of regions from one another by the creation of geographic barriers. As these barriers were produced by geological or climatic processes, cladogram topology of the vicariance tree can be used to infer the extent and relative timing of geological or climatic processes.

As discussed in Chapters 5 and 6, there are several geological processes that can produce vicariance, but there are two primary ones: continental rifting and changes in relative sea level. Continental rifting would influence marine species that live near the coastline or terrestrial species that occupy the interior of continents. The expectation is that both types of organisms generally react in a similar way to such a tectonic event. As a formerly continuous patch of land or seashore, within the zone of rifting divides into two or more separate regions, the originally contiguous populations become separated and eventually may diverge evolutionarily and speciate.

Throughout Earth history sea level, relative to the position of the continents, has risen and fallen (e.g., Hallam, 1992), and such oscillations can be related to both plate tectonic and climatic events. Changes in sea level also influence the distribution of terrestrial and marine species, albeit differently. In terrestrial species, an increase in sea level can divide populations that are on either side of the rising waters, so a rise in sea level can encourage vicariance in terrestrial organisms. However with marine species, a rise in sea level might join what were formerly isolated marine species by removing a geographic barrier, such as a stretch of dry land, that had separated them, which corresponds to geodispersal. By contrast, a fall in sea level will generally lead to vicariance in marine species. As sea level is lowered, populations are likely to become isolated by emergent spits of land, which would encourage vicariant differentiation.

Thus, a vicariance tree can be considered a statement about the relative timing of geological and climatic processes that influence patterns of evolution and distribution. Depending on the types of organisms considered, marine or terrestrial, we can look at each divergence point between different areas on the vicariance tree as the formation of a barrier due either to rifting continents or sea-level rise or fall. This information can often be used to make predictions. For example, there is a large and diverse literature in the field of geology that concentrates on reconstructing the timing of plate tectonic events (see, e.g., Scotese and McKerrow, 1990). The vicariance tree that results from a biogeographic analysis can be viewed as an independent biological test of different models of how plates and thus biotas may have separated from one another. These results about the timing of the separation of continents can be compared with the results gathered by geologists working in the area of paleomagnetism, or geochemists applying geochronological techniques, or structural geologists looking at geological provinces spanning now disparate sites, or field geologists looking at packages of strata. Thus, paleobiogeographic analysis is an important technique that paleontologists can apply to the study of geological problems.

Moreover, the vicariance tree can tell us something more about the nature of the evolutionary process in general. If such trees show little or no resolution among the areas, it means that the different clades analyzed had largely individualistic biogeographic patterns. This would imply that Earth history events at this time or in these regions did not powerfully influence patterns of evolution and geographic distribution. By contrast, if a well-resolved vicariance tree results, we can be confident that Earth history events played a prominent role in influencing patterns of evolution and distribution in these clades. Referring again to the discussion in Chapters 5 and 6, we note that these divergent patterns would imply a very different evolutionary perspective. They define how the geological world impinges on the biological world, and to what extent the Earth and its biota have coevolved. Such coevolution implies that the evolutionary process in general is largely contingent, historical, and unpredictable, and further indicates that there should be pulsed times of major evolutionary change that occur in several groups and which are related to some external forcing agent. This view is akin to that expressed in publications such as Matthew (1939), Eldredge and Gould (1972), Eldredge (1979, 1985a), Gould (1980, 1989), Vrba (1980, 1985), and Brooks and McLennan (1991).

9.8.4.7. Interpreting the Geodispersal Tree

The geodispersal tree also provides important biogeographic information, about the relative time that geographic barriers between regions fell, joining formerly separated regions and biotas and causing geodispersal. Areas that share a recent common ancestor or node on the area cladogram were fairly recently joined as a single area and the barriers that fell most recently were those that separated these regions from one another. Areas that are more distantly related on the tree were joined as a single area more distantly in the past, and the barriers that divided them fell much earlier. In this way, area cladogram topology can be used to infer the timing of the merging or joining of regions with one another owing to the fall of geographic barriers. The processes that remove or efface these barriers are produced by geological or climatic events, just as was the case with the processes that cause them to form. Thus, as with the vicariance tree, the topology of the geodispersal tree can be used to infer the extent and timing of geological processes.

Basically, the same types of geological processes that cause vicariance can cause geodispersal. One geological process that might produce geodispersal by joining formerly separated regions with isolated biotas is continental collision. This process would influence marine species that live near the coastline or terrestrial species that occupy the interior of continents. Both types of organisms are predicted to generally react similarly to such a tectonic event. In this case, two or more formerly separated regions of land and seashore will come to be joined into a single region. As the originally separated species become contiguous they may expand their ranges into the newly conjoined land mass or seashore. Although continental collisions will initially produce conditions ripe for geodispersal, after extended periods of tectonic interaction, these collisions may actually contribute to patterns of vicariance. (Recall, from Chapter 6, range expansion requires subsequent vicariance to make geodispersal truly macroevolutionary in scope.) For example, the Himalayas are the long-term by-product of a Cenozoic collision between India and Asia. The Himalayas now represent a major barrier to the northward movement of Indian terrestrial species and the southward movement of Asian terrestrial species. Soon after the collision, Asian species could have migrated into India and vice versa. However, as the Himalayas were built up, populations of species would have become isolated on either side of the range, leading to vicariance.

Changes in relative sea level can also cause geodispersal. A fall in sea level will allow formerly separated terrestrial biotas to become contiguous, and then terrestrial species originally confined to narrower regions can expand their range to encompass the larger now joined regions. Thus, a fall in sea level can often encourage geodispersal in terrestrial organisms. Geodispersal in marine organisms could be produced by a rise in sea level that would cover land that had intervened between taxa. The removal of these barriers would allow species to expand their range so that they come to occupy new, broader areas, thus changing distributional patterns.

Thus, a geodispersal tree, like a vicariance tree, can be considered a statement about the relative timing of geological and climatic processes that influence patterns of evolution and distribution. Again, depending on the types of organisms considered, marine versus terrestrial, we can look at each divergence point on the tree between different areas as the fall of a barrier due to either continental collision joining plates or sea-level rise or fall. This information can be used to make predictions about the sequence of geological events. Finally, just as was the case with the vicariance tree, the geodispersal tree, if well resolved, indicates that Earth history events played a major role in influencing biogeographic patterns. A poorly resolved geodispersal tree indicates that clades responded individualistically to a myriad of Earth history effects or that their range expansion was governed largely by ecological factors.

9.8.4.8. Comparing the Vicariance and Geodispersal Trees

There is one additional type of biogeographic information that can come from applying this modified version of BPA. The patterns in the vicariance tree can be compared with the patterns in the geodispersal tree, which will tell us something about the nature of the processes causing the biogeographic patterns. Comparisons emphasize the search for congruence between the two trees. Strong similarities between them indicate that the same processes that caused vicariance also caused geodispersal because each tree can be thought of as providing information about the relative time that barriers formed. The geodispersal tree indicates the most recent barriers to fall; the vicariance tree indicates the most recent barriers to rise. Similarities between the trees imply that the most recent barriers to rise were also the most recent ones to fall. This can happen when vicariance and geodispersal occur in an oscillating sequence to repeatedly separate and reunite the same regions, which will result from the action of cyclical geological or climatic processes that create and then efface geographic barriers.

Sea level rises and falls on time scales commensurate with speciation in most organisms. Focusing on marine organisms, one can see how lowering the sea level might isolate populations across newly emergent geographic barriers. Eventually these populations would speciate, producing two closely related species separated by a barrier. If sea level were to rise again to a height sufficient to breach these topographic barriers then the newly formed species could expand their ranges back into the region where their ancestors had been distributed.

If extensive similarity between patterns of vicariance and geodispersal is uncovered then one could conclude that cyclical geological or climatic processes played a strong role in influencing biogeographic patterns in the regions being studied. By contrast, if differences emerge between the two trees then one might conclude that cyclical processes had less influence on biogeographic patterns and that geological or climatic processes that are not cyclical, or at least not cyclical on a timescale commensurate with speciation, motivated the biogeographic patterns. These kinds of geological processes might involve events such as collisions between continents, which only allow for a single instance of range expansion. When patterns of vicariance and geodispersal diverge it is also possible that the biogeographic patterns might not be caused either by geological processes or climatic processes but rather by a single traditional dispersal event. After this event, there would be little possibility for subsequent dispersal back into the region from whence the taxa originated and thus no repeating patterns of range expansion and vicariance.

Techniques developed by Simberloff (1987) and Siddall (1996) make it possible to quantify similarities between geodispersal and vicariance trees. These techniques pose a question as to what the probability that the same tree could arise by chance. If that probability is low, one can be confident of the results.

9.8.5. Results from BPA: A Reiteration

One can think of the modified version of BPA described here [see also Lieberman and Eldredge (1996) and Lieberman (1997)] as providing several different types of information that are relevant to any biogeographic study. First, the output from BPA can help quantify the extent to which Earth history events have controlled the evolutionary process. The role that these events play in driving evolution has important philosophical and theoretical implications for our understanding of the history of life, evolutionary biology, and paleontology. Further, this version of BPA can be used to indicate sequences of Earth history events, be they tectonic or climatic, and this information has direct relevance to basic research in geology. Finally, the relative role that cyclical geological or climatic processes play in influencing biogeographic patterns can also be quantified.

9.9. Arguments about Using Parsimony Algorithms in Biogeography

I noted earlier that PAUP 4.0 (Swofford, 1998), the program used to evaluate BPA matrices, is a parsimony-based algorithm. When used in phylogenetic studies it produces the best-supported tree or trees based on the character data, be they morphological or molecular, that are input into the data matrix. These data involve the distribution (presence or absence) of characters in the different taxa being studied. The tree of the shortest length that invokes the fewest number of independent evolutionary acquisitions of characters is the one that is preferred. As discussed in Chapter 8, the use of parsimony as a general criterion in phylogenetic studies has been discussed and validated by numerous authors including Platnick and Gaffney (1977, 1978a,b), Gaffney (1979), Eldredge and Cracraft (1980), Nelson and Platnick (1981), Wiley (1981), Sober (1988), and Brooks and McLennan (1991). It appears to have overwhelming philosophical and methodological support as a general method for choosing the best-supported pattern of evolutionary relationship.

The character data in biogeographic studies involve the regions that taxa occur in and how geographic distributions changed as the group evolved. When a parsimony algorithm is used in biogeographic studies, it determines the biogeographic pattern that is best supported by these data, and the one most frequently replicated in the character data is the one that will be retrieved by analysis. Parsimony is a method of choosing among many competing alternative biogeographic patterns; we are trying to choose the overwhelming pattern without invoking additional biogeographic events for which there is no evidence.

Again, there is an analogy with phylogenetic analysis. There, the goal was to minimize the number of times that a character evolved independently. The possession of the same feature was treated, at least initially, as evidence for the fact that two taxa were closely related. The same is true in biogeography. What we are trying to minimize is the number of times that a taxon (an ancestral node or an actual terminal taxon) assumed the same geographic state. It is conceivable that even though one finds an ancestor and its descendant distributed in the same two areas, at some time before the ancestor diverged into its descendant it may have dispersed back and forth between other regions. However, there is no evidence to support this; it is not a hypothesis that is directly evident from inspection of the data. BPA invokes parsimony to minimize unsubstantiated biogeographic events for which there are no data. Brooks (1985, 1988, 1990), Kluge (1988), Mayden (1988), Wiley (1988*a,b*), Funk and Brooks (1990), Brooks and McLennan (1991), Lieberman and Eldredge (1996), and Lieberman (1997) have all strongly endorsed the use of parsimony-based algorithms in biogeographic analysis.

Although parsimony at first glance seems a valuable tool in biogeographic analysis, its use has not been universally endorsed. For example, Cracraft (1988) argued that although parsimony can reasonably be applied in phylogenetic studies of organisms it should not be used in biogeographic studies. He argued that phylogenetic and biogeographic studies are different because evolution occurs by common descent such that the taxa and the characters they possess have a single evolutionary history. By contrast, he claimed that areas are capable of having multiple cladistic histories.

In a superficial sense Cracraft (1988) is correct in saying that areas can have multiple cladistic histories, but in a deeper sense his reasoning is flawed because his statement does not mean that biogeographic analysis is not analogous to phylogenetic analysis. What he meant by multiple cladistic histories in biogeography is that areas may show episodes of range expansion followed by vicariance followed by range expansion. Therefore, areas are not strictly monophyletic since they can have histories that indicate independent derivations of the taxa they contain from several different regions or at several different times. These phenomena are equivalent to divergence and reticulation. However, species are taxa, which can show the same type of divergence and reticulation. As discussed in Chapter 7, areas are really analogous to species, and it has not been suggested that species cannot be subjected to phylogenetic analysis.

The fact that tokogenetic and phylogenetic types of relationships may prevail among areas does not preclude the use of parsimony-based techniques to study these patterns of area relationship any more than the nature of species precludes their use to study how species evolve. In biogeography, these twin patterns of relationship, engendered by geodispersal and vicariance, do create complications; however, these can be dealt with by expanding the way biogeographic patterns are analyzed, while still using a parsimony-based approach.

Sober (1988) also had reservations about the use of parsimony to study biogeographic patterns. His criticism centered on the fact that traditionally in vicariance biogeography vicariance has been treated as a fundamentally different type of explanation than dispersal, the latter being treated as noise obscuring the pattern of interest. Vicariance biogeographers have related vicariance to dispersal in the same way that homology is distinguished from homoplasy in phylogenetic systematics, i.e., the former reflects the underlying signal of patterns of evolutionary relationship whereas the latter is noise.

Sober (1988) argued that "(t)here is no abstract and general argument for favoring vicariance over dispersal that flows from the same source that leads us to prefer homology over homoplasy" (Sober, 1988, p. 252–253), and I concur with his statement. As I have been arguing throughout this book, there can be types of range expansion that are directly analogous to vicariance because they can lead to congruence in different clades. We do not need a biogeographic method that minimizes range expansion relative to vicariance. Rather, we need a method that first searches for the best-supported pattern of vicariance in the data and then for the best-supported pattern of geodispersal. The manner in which parsimony has been employed in vicariance biogeography thus far is spurious because as Sober (1988) correctly pointed out it pits vicariance against dispersal. However, this does not mean that parsimony-based methods can never be used in historical biogeography. Instead, we have to make sure that what is being minimized is the right type of noise. Thus, while his criticisms of the traditional use of parsimony in vicariance biogeographic studies are valid, it does not *per se* represent an indictment against the use of any parsimony criterion in historical biogeography.

Other criticisms of the use of parsimony in historical biogeography are basically along the same lines as those of Cracraft (1988). Therefore, for the same reasons elucidated above, they are not valid, but the interested reader should refer to Page (1990) and the discussions in Morrone and Carpenter (1994) and Morrone and Crisci (1995) for additional amplification. In the end, parsimony is the best way to choose among several different competing hypotheses. These algorithms accept the data at face value. By contrast, at least at this time, designing algorithms based on probabilities of movement and allopatric speciation would be so divorced from the actual hard data and so devoid of factual substantiation that they would have to be viewed solely as within the realm of fantasy.

9.10. Other Criticisms of Brooks Parsimony Analysis that Are No Longer Valid

Apart from the invalid criticism concerning BPA's use of parsimony to study biogeographic patterns, there have been some other questions raised about BPA. Although these reservations are also without merit they are worth discussing. For example, articles by Morrone and Carpenter (1994) and Morrone and Crisci (1995), which dealt with what the authors perceived as problems with BPA, amplified the concerns of certain other authors and focused on interpreting its results. (Their discussion relates to BPA as traditionally implemented, not to the modified version outlined here.)

After an area cladogram is generated, there will be a set of biogeographic characters that support that tree. This is akin to phylogenetic analysis, when there is a set of morphological or molecular characters that support an evolutionary tree. After a phylogenetic analysis is conducted, the computer packages PAUP (Swofford, 1998) and MacClade (Maddison and Maddison, 1992) allow one to see which characters support different parts of the tree, which is often referred to as mapping the characters to the tree.

Morrone and Carpenter (1994) and Morrone and Crisci (1995) discussed what happened when biogeographic characters were mapped to the tree after BPA was applied. In traditional BPA, the characters that support the final area cladogram include the ancestral nodes and the terminal taxa of each of the individual clades considered in biogeographic analysis. These characters can be subsequently mapped onto the final area cladogram using a parsimony algorithm, the basic method also used to map characters to a tree in phylogenetic analysis. They showed that when this is done sometimes the ancestral nodes mapped onto the tree much earlier, or even on different branches, than their supposed descendants. They argued that this was highly paradoxical because it implied disassociation between ancestors and descendants. Because of this inconsistency they suggested that the method lacked general validity. This problem was acknowledged by Brooks (1990), and he did discuss various modifications by which characters could be optimized to trees after a BPA is performed, which blunted the initial criticism, but in spite of these modifications, Morrone and Carpenter (1994) and Morrone and Crisci (1995) still strongly criticized BPA's utility on these grounds.

Although the weaknesses pointed out by these critics may or may not be serious they are caused by a particular procedure in BPA, and the problem is obviated once the procedure used in the modified version is implemented. The problem is related to the way in which the biogeographic states of terminal taxa are placed at the ancestral nodes during the generation of the individual area cladograms. In standard BPA the ancestral nodes of the individual area cladograms are optimized using inclusive OR-ing. Basically as one steps down the tree a new area is added to each of the ancestral nodes reflecting the states of the terminal taxa (see discussion and Fig. 19). This procedure of optimizing characters is not a most-parsimonious solution, and in fact it is this optimization procedure that leads to the problem that caused Morrone and Carpenter (1994) and Morrone and Crisci (1995) to criticize the technique. It is not surprising that these unparsimonious optimizations should conflict later with the parsimonious optimizations of the characters onto the final area cladogram. It was for this reason, and also because there are some other problems associated with the inclusive OR-ing method, that we (Lieberman and Eldredge, 1996) argued that a parsimony-based algorithm such as Fitch parsimony should be used to determine the ancestral nodes for each of the individual cladograms considered in biogeographic analysis. When such a parsimony-based procedure is used initially, the inconsistencies at issue here are ameliorated, and the criticisms raised about the character optimizations are not sufficient to cause serious concern about the use of BPA; however, even if they were, the modifications to BPA that we have proposed (Lieberman and Eldredge, 1996 and herein) are enough to eliminate these concerns.

9.11. Parsimony Analysis of Endemicity

In this discussion of phylogenetic approaches to biogeographic analysis, one method originally used specifically for the study of fossil taxa is worthy of mention. Referred to as parsimony analysis of endemicity (PAE), it was developed by Rosen (1988). In PAE, a data matrix is generated in a fashion that is in some ways akin to the way BPA works. The rows of the matrix have the geographic regions of interest. Then, distributional data for taxa in these regions are recovered. Taxa are treated as either present in, or absent from, a region. Presence is scored with a "1," absence with a "0," and an all "0" outgroup is added as one of the rows. Then the data matrix is analyzed using a parsimony-based algorithm such as PAUP (Swofford, 1998), the justification for its use in PAE being the same as for BPA.

The result of a PAE is an area cladogram depicting patterns of relationship among areas. Fortey and Cocks (1992) have demonstrated how this method can be successfully applied to the analysis of paleobiogeographic patterns, and Raxworthy and Nussbaum (1996) and Emerson *et al.* (1997) have demonstrated its utility when it is applied to the study of extant biogeographic patterns. This method is useful when something is known about the distribution of taxa in regions that are being studied for biogeographic patterns, but a broader understanding of phylogenetic patterns within these taxa is lacking. Thus, its results should be treated as good first approximations. However, by only incorporating information about the distribution of taxa and not considering their evolutionary relationships, PAE, like the phenetic approaches described above, cannot capture information about how geographic distributions of groups have changed as these groups have evolved. Thus, there are additional data that can be mined by biogeographic methods, when phylogenetic relationships are understood, that PAE cannot capture.

9.12. Case Studies Assessing the Efficacy of Components Analysis vs. Brooks Parsimony Analysis that Used the Extant Biota

Brooks (1985, 1988), Kluge (1988), Mayden (1988), Wiley (1988*a,b*), Funk and Brooks (1990), Brooks and McLennan (1991), and Wiley *et al.* (1991) discussed the strengths of BPA as a general method for biogeographic analysis at length, and presented several case studies demonstrating its validity. There have also been new studies since the publication of these works, that continue to demonstrate its strength as a general method of biogeographic analysis. All of these studies concentrated on BPA as traditionally implemented, and traditional BPA is not designed to deal with congruent episodes of range expansion. It has been argued herein that these represent an important type of biogeographic process that needs to be considered. However, the fact that BPA performed extremely well in biogeographic analysis, as well as or better than other biogeographic methods, in spite of the fact that it did not consider geodispersal demonstrates its great strength and resiliency as a general discovery procedure in biogeography.

An example of one of these recent studies is the work of Morrone and Carpenter (1994), who considered how well different methods of biogeographic analysis worked. Although these authors criticized aspects of BPA, they actually found that it was as good or better than components analysis as a biogeographic method, and significantly better than less well-established methods of biogeographic analysis such as three area statements. In another study Enghoff (1995) analyzed the historical biogeography of the Arctic. Although his study did not explicitly evaluate BPA relative to other biogeographic methods, he compared some of the assumptions of BPA with those of components analysis; specifically, he compared assumption 0 with assumptions 1 and 2. Assumption 0 treats the presence of a taxon or ancestral node in more than one area as evidence that these areas shared a common geological history. Assumptions 1 and 2 were procedures implemented in components analysis that involved altering the distribution of some but not all aspects of tree topology, and earlier in this chapter these assumptions were questioned on several grounds. Enghoff (1995) endorsed assumption 0 both on theoretical grounds and because it performed better in recovering biogeographic signal. This endorsement of assumption 0 can be treated as an overall endorsement of the methodological approach taken in BPA. Enghoff (1996) again considered assumption 0 and compared it with assumption 2. He found that at least in the case of widespread taxa the use of assumption 0 had significant strengths over the use of assumption 2, although he did not endorse its use in all cases. His study can be treated as providing further, but not unequivocal support for the assumptions underlying traditional BPA. Finally, recent work of Soest and Hajdu (1997) indicated that BPA and components analysis performed equally well in recovering biogeographic patterns in marine sponges.

9.13. Paleobiogeographic Studies Using Phylogenetic Approaches and the Modified Version of Brooks Parsimony Analysis

There have been few studies that have applied cladistic biogeographic techniques to the fossil record. One, Fortey and Cocks (1992), utilized PAE and was described above. The only other cladistic paleobiogeographic studies that have been published are ours (Lieberman and Eldredge, 1996; and Lieberman, 1997). Each of these studies utilized the modified version of BPA discussed herein. These are worth considering in greater depth because they show the strength of this analytical approach, while illustrating the range of questions that it can be used to answer. Each study used phylogenetic and distributional data from different clades of trilobites.

9.13.1. Cladistic Biogeography of Middle Devonian Trilobites

We studied (Lieberman and Eldredge, 1996) biogeographic patterns in trilobites during the Middle Devonian, roughly 380 million years ago. This was a geologically complex period during which continents were coming together and colliding with one another (Fig. 8). In eastern North America these collisions led to an incipient mountain range that was the geological precursor to the Appalachian Mountains, and this time of mountain-building is referred to as the Acadian Orogeny. Collisions amalgamate formerly separated regions, effacing geographic barriers between taxa, but they can also eventually lead to the development of major geographic barriers between regions, such as mountain ranges. During the Middle Devonian there were also several major episodes of global sea-level rise and fall on the order of tens or even hundreds of meters, which may have been partly related to the tectonic events. Such profound changes in sea level will first form and later eliminate geographic barriers for benthic marine organisms such as trilobites. It can clearly be seen how these overlaid geological and climatic processes would powerfully influence patterns of evolution and distribution by causing vicariance and geodispersal.

We investigated (Lieberman, 1994; Lieberman and Kloc, 1997) phylogenetic patterns in five clades of trilobites that were distributed across many different regions during the Middle Devonian. The regions chosen, illustrated in Figs. 8 and 11, correspond to areas of endemism as defined in Chapter 7: they contained large numbers of unique taxa and were defined spatially by major geographic barriers. We (Lieberman and Eldredge, 1996) converted each of the five phylogenies to area cladograms by substituting the geographic occurrence of each species for the species name. These occurrences were then optimized to the ancestral nodes using the implementation of the Fitch (1971) algorithm described above. The converted area cladograms from each of these phylogenies are shown in Figs. 20 and 21.

The information from each of the area cladograms was coded, as described above, into two BPA matrices: a vicariance matrix and a geodispersal matrix, which are given in Tables 5 and 6, respectively. Then both matrices were analyzed using the program PAUP 3.0q (Swofford, 1993). The resulting most-parsimonious area cladograms, the vicariance and geodispersal trees, are shown side by side in Fig. 22. Various tests, described in Lieberman and Eldredge (1996), revealed that each of these area cladograms was well supported, implying that we can have confidence in the biogeographic patterns they predict. Each cladogram indicates that eastern North America (ENA) is a well-supported biogeographic region and that the Canadian Arctic and Armorica, a region comprising Central Europe, the Iberian Peninsula, and northern Africa also share strong biogeographic affinity.

Now can we relate these statements about biogeographic affinity to a sequence of tectonic events, demonstrating the important impact that Earth history events have on evolution? The answer is yes, and the clearest example of this that emerged from our study (Lieberman and Eldredge, 1996) involved

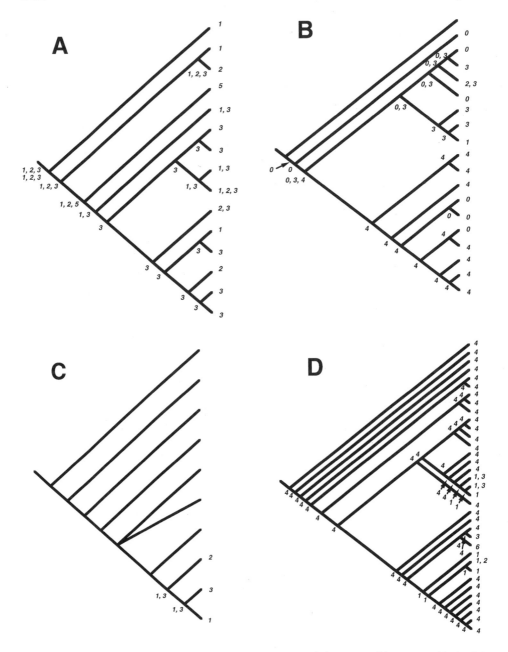

FIGURE 21. Area cladograms from four trilobite clades with biogeographic states optimized to nodes for (A) the genus *Crassiproetus*, (B) the genus *Dechenella*, (C) the "*Thebanaspis* clade," and (D) the Asteropyginae. 0–4 given in caption of Fig. 20, 5 = Kazakhstan, and 6 = northern South America. From Lieberman and Eldredge (1996), used by permission.

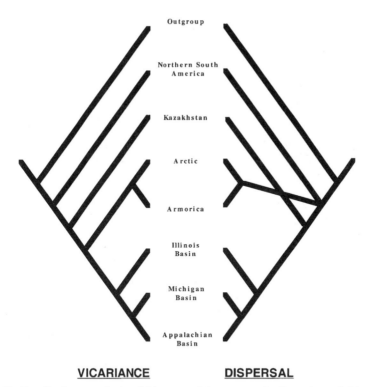

FIGURE 22. Results from analysis of biogeographic patterns in Devonian trilobites based on Lieberman and Eldredge (1996). The vicariance tree is on the left, and the geodispersal tree is on the right. There is a high degree of similarity between the vicariance and geodispersal trees, and the geodispersal tree is well resolved. From Lieberman (1999a), used by permission.

the different basins of ENA: the Appalachian, Michigan, and Illinois basins. These basins were topographic depressions that were marine environments throughout most of the Devonian period. Their formation and growth were related to a series of tectonic events including the Acadian Orogeny. That orogeny was driven by collisions between ENA and either Armorica or some microcontinents that became attached to North America (Ettensohn, 1985; Kent, 1985; Soper et al., 1992). These collisions lead to buckling within ENA that helped accentuate the geographic barriers separating the different basins in that region. Principally, the geographic barriers are large arches that were at various times above sea-level and thus would have prevented the direct movement of marine organisms between the basins.

The vicariance tree made some statements about the timing of the emergence of barriers that separated the three basins in ENA. It indicated that the Illinois Basin became separated from both the Appalachian and Michigan basins first, and then the latter two were subsequently separated from one another.

Beaumont et al. (1988) produced a geophysical model of how the Acadian Orogeny might have influenced the geometries of the basins in ENA,

Table 5. Modified BPA Coding of the Vicariance Matrix for the Area Cladograms Shown in Figs. 20 and 21[a]

	12345	678901	11111 12345	11112 67890	22222 12345	22223 67890	33333 12345
Ancestor	00000	00000	00000	00000	00000	00000	00000
Arctic	11110	00000	00000	00000	00000	00000	00000
Appalachian Basin	11121	11111	11211	12111	00000	00011	21121
Illinois Basin	11110	10000	00011	12211	00000	00011	11112
Michigan Basin	11121	12111	11111	11122	11111	11111	11111
Armorica	01210	00000	00000	00000	00000	00000	00000
Kazakhstan	00000	00000	00000	00000	00000	00000	00000
N. South America	00000	00000	00000	00000	00000	00000	00000

	33334 67890	44444 12345	44445 67890	55555 12345	55556 67890	66666 12345	66667 67890
Ancestor	00000	00000	00000	00000	00000	00000	00000
Arctic	00000	00000	00000	00000	00011	11211	11000
Appalachian Basin	11211	00001	11000	10000	00000	00000	00000
Illinois Basin	11100	00000	01010	00011	00000	00000	00000
Michigan Basin	00212	11111	11111	01100	11100	01211	12111
Armorica	00000	00000	00000	00000	00000	01100	00000
Kazakhstan	12100	00000	00000	00000	00000	00000	00000
N. South America	00000	00000	00000	00000	00000	00000	00000

	77777 12345	77778 67890	88888 12345	88889 67890	99999 12345	999901 67890	11111 00000 12345
Ancestor	00000	00000	00000	00000	00000	00000	00000
Arctic	00212	10000	00111	00010	00000	00000	00000
Appalachian Basin	10000	00000	00000	00000	00001	01120	00000
Illinois Basin	01000	00000	00000	00000	00000	10000	00000
Michigan Basin	01121	10000	00000	00000	00001	01210	00000
Armorica	00000	21111	11000	11101	11110	00001	11111
Kazakhstan	00000	00000	00000	00000	00000	00000	00000
N. South America	00000	00000	00000	00000	00000	00000	00000

	11111 00001 67890	11111 11111 12345	11111 11112 67890	11111 22222 12345	11111 22223 67890	11111 33333 12345	11111 33334 67890
Ancestor	00000	00000	00000	00000	00000	00000	00000
Arctic	00000	00000	00000	00000	00000	00000	00000
Appalachian Basin	00000	00000	00000	00000	00000	00001	11110
Illinois Basin	00000	00000	00000	00000	00000	00000	00000
Michigan Basin	00000	00000	00000	00000	00000	00000	00110
Armorica	11111	11111	11111	11111	11111	11110	00001
Kazakhstan	00000	00000	00000	00000	00000	00000	00000
N. South America	00000	00000	00000	00000	00000	00000	00000

	11111 44444 12345	11111 44445 67890	11111 55555 12345	11111 55556 67890	11111 66666 12345	111 666 678

Ancestor	00000	00000	00000	00000	00000	000
Arctic	00000	00000	00000	00000	00000	000
Appalachian Basin	00000	00001	11111	00000	00000	000
Illinois Basin	00000	00100	00001	00000	00000	000
Michigan Basin	00000	00000	00000	00000	00000	000
Armorica	11111	11000	00000	11111	11111	111
Kazakhstan	00000	00000	00000	00000	00000	000
N. South America	00000	00010	00000	00000	00000	000

"0" represents absence from a region; "1" and "2" represent presence in a region. Ancestor refers to the ancestral biogeographic condition for the clades considered. Multistate characters are ordered (additive). Characters 1–28 are the codings from the phylogeny of *Basidechenella*; characters 29–58 from the phylogeny of *Crassiproetus*; characters 59–94 from the phylogeny of *Dechenella*; characters 95–99 from the phylogeny of the "*Thebanaspis clade*"; and characters 100–168 from the phylogeny of the Asteropyginae.

specifically, the manner and timing in which they would have become separated, and the modeled pattern matched that recovered from the patterns of trilobite distribution and evolution. Here is a clear case where the tectonic events contributed significantly to patterns of speciation in a group of marine organisms, and this is further evidence in support of the proposition of Croizat (1958, 1964) that the Earth and its biota have coevolved. It also indicates that evolution is very much a contingent process, driven by external events (Gould, 1989).

Additional information about the role of Earth history phenomena in governing the evolutionary process can be extracted from this case study by comparing the vicariance and geodispersal trees (see Fig. 22). It can be readily seen that these trees are very similar. In particular, they both predict a close relationship among the different basins of ENA, with the Appalachian and Michigan basins sharing a more recent relationship than either shares with the Illinois basin, and they also both predict a close biogeographic relationship between the Canadian Arctic and Armorica. They do, however, diverge in a few instances, particularly concerning the relative position of Kazakhstan.

First, let us focus on the similarities. For illustrative purposes I will concentrate on the different basins of ENA. Each tree makes some statement about geographic barriers. The vicariance tree predicts the relative time that geographic barriers emerged, and the geodispersal tree predicts the relative time that they fell. In this case, the biogeographic data indicate that the relative time that the barriers between the Appalachian, Michigan, and Illinois basins formed corresponds to the relative time that the barriers between these regions were removed. For example, the vicariance tree predicted that the barrier separating the Illinois Basin from the others was the first to emerge and the geodispersal tree predicted that the barrier separating the Illinois Basin from the other basins was the first to fall.

As these biogeographic patterns emerged from the synthesis of evolutionary patterns in several independent clades and are well supported by various tests of tree support [described in Lieberman and Eldredge (1996)], the similarities between the vicariance and geodispersal trees are best interpreted as implying that they were governed by cyclical geological and climatic processes. These operated first to create barriers to the free movement of

Table 6. Modified BPA Coding of the Geodispersal Matrix for the Area Cladograms Shown in Figs. 20 and 21[a]

	12345	678901	11111 12345	11112 67890	22222 12345	22223 67890	33333 12345
Ancestor	00000	00000	00000	00000	00000	00000	00000
Arctic	11000	00000	00000	00000	00000	00000	00000
Appalachian Basin	11011	10111	11111	11000	00000	00011	11110
Illinois Basin	11000	20000	00021	11010	00000	00011	01101
Michigan Basin	11011	11111	10111	01101	11111	11111	01100
Armorica	02100	00000	00000	00000	00000	00000	00000
Kazakhstan	00000	00000	00000	00000	00000	00000	00000
N. South America	00000	00000	00000	00000	00000	00000	00000

	33334 67890	44444 12345	44445 67890	55555 12345	55556 67890	66666 12345	66667 67890
Ancestor	00000	00000	00000	00000	00000	00000	00000
Arctic	00000	00000	00000	00000	00011	11100	00011
Appalachian Basin	10110	00002	11000	02000	00000	00000	00200
Illinois Basin	10100	0000	02020	00002	00000	00000	00000
Michigan Basin	00211	11111	11111	11111	11100	02111	11110
Armorica	00000	00000	00000	00000	00000	02000	00000
Kazakhstan	21100	00000	00000	00000	00000	00000	00000
N. South America	00000	00000	00000	00000	00000	00000	00000

	77777 12345	77778 67890	88888 12345	88889 67890	99999 12345	999901 67890	11111 00000 12345
Ancestor	00000	00000	00000	00000	00000	00000	00000
Arctic	11101	00000	00211	00200	00000	00000	00000
Appalachian Basin	00000	00000	00000	00000	00001	11010	00000
Illinois Basin	02000	00000	00000	00000	00000	20000	00000
Michigan Basin	11110	00000	00000	00000	00001	11100	00000
Armorica	00000	11111	11100	11111	11110	00001	11111
Kazakhstan	00000	00000	00000	00000	00000	00000	00000
N. South America	00000	00000	00000	00000	00000	00000	00000

	11111 00001 67890	11111 11111 12345	11111 11112 67890	11111 22222 12345	11111 22223 67890	11111 33333 12345	11111 33334 67890
Ancestor	00000	00000	00000	00000	00000	00000	00000
Arctic	00000	00000	00000	00000	00000	00000	00000
Appalachian Basin	00000	00000	00000	00000	00211	11000	00000
Illinois Basin	00000	00000	00000	00000	00000	00000	00000
Michigan Basin	00000	00000	00000	00000	00020	20000	00000
Armorica	11111	11111	11111	11111	11100	00111	11111
Kazakhstan	00000	00000	00000	00000	00000	00000	00000
N. South America	00000	00000	00000	00000	00000	00000	00000

	11111 44444 12345	11111 44445 67890	11111 55555 12345	11111 55556 67890	11111 66666 12345	111 666 678

Ancestor	00000	00000	00000	00000	00000	000
Arctic	00000	00000	00000	00000	00000	000
Appalachian Basin	00000	00002	11111	10000	00000	000
Illinois Basin	00000	00000	00020	00000	00000	000
Michigan Basin	00000	00020	00000	00000	00000	000
Armorica	11111	11111	00000	21111	11111	111
Kazakhstan	00000	00000	00000	00000	00000	000
N. South America	00000	20000	00000	00000	00000	000

"0" represents absence from a region; "1" and "2" represent presence in a region. Ancestor refers to the ancestral biogeographic condition for the clades considered. Multistate characters are ordered (additive). Characters 1–28 are the codings from the phylogeny of *Basidechenella*; characters 29–58 from the phylogeny of *Crassiproetus*; characters 59–94 from the phylogeny of *Dechenella*; characters 95–99 from the phylogeny of the "*Thebanaspis* clade"; and characters 100–168 from the phylogeny of the Asteropyginae.

marine organisms and later to remove them. As mentioned above, during the Middle Devonian there were several major episodes of sea-level rise and fall. When sea level fell, the arches that separated the different basins in ENA from one another became emergent and served as profound barriers to the free movement of trilobite taxa. This would serve to isolate what had once been continuously distributed species, promoting speciation. When sea level rose these arches would sometimes have been breached, allowing species formerly isolated in a single basin to expand their range to include the other basins. How this would happen is illustrated in Fig. 23. Thus, we have a clear case in which biogeographic patterns indicate that not only did Earth history factors play a prominent role in influencing patterns of evolution, but they also operated in a cyclical fashion. (Whether one treats these sea-level changes as climatic or geological is an interesting topic that may be open to several interpretations; however, in either event the role of Earth history in promoting evolution is clear.)

There are other paleontological studies that have demonstrated the prominent role that cyclical Earth history events play in promoting large-scale patterns of evolution. Among the studies that immediately comes to mind are those of Vrba (1980, 1985, 1992, 1996), in which she put forward evidence for her Turnover-Pulse hypothesis. In these papers she considered the geographic ranges of many tropical mammal species distributed in the Neogene fossil record of South America and Africa, which contracted and expanded, in response to climatic cycles of cooling and warming, respectively. As temperatures cooled, the climatic changes led to contractions of the preferred habitats of these tropical mammals, and thus contractions in their geographic ranges, which promoted the isolation of populations of these species. This led subsequently to pulsed speciation and extinction events in many different groups, from whence the name of the hypothesis is derived. When temperatures later rose, these newly evolved taxa could expand their ranges. It is highly likely that the evolution of our own lineage, the genus *Homo*, was driven by these pulsed climatic changes. Such studies could also easily be integrated into a cladistic biogeographic context such as the one described herein. It would be predicted that the clades affected by these

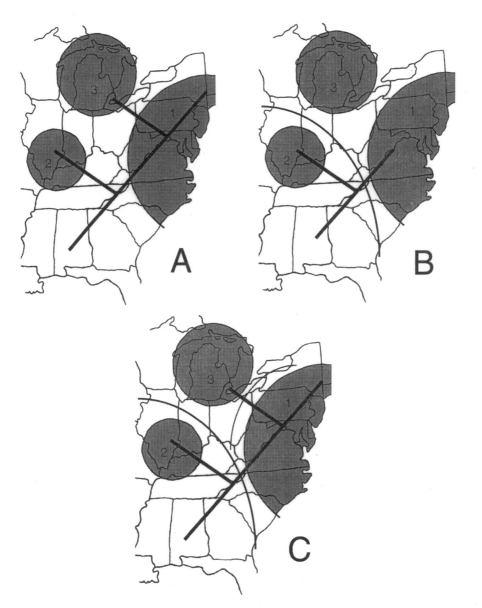

FIGURE 23. Hypothetical scenario depicting how cyclical climatic processes such as sea-level rise and fall can produce similar patterns of vicariance and geodispersal: (A) Devonian biogeographic patterns for the different basins of ENA predicted by both the geodispersal and vicariance trees, during a period of relatively high sea level no prominent barriers separated these basins and a species of marine invertebrate was present in all of them. (B) After sea level fell, the topographic arch separating the Illinois Basin from the others became an emergent barrier preventing the movement of marine organisms; populations of that species became isolated on either side of the barrier, diverged, and then speciated, and a phylogeny relating them is shown overlying the basins. (C) sea-level fell even further, the topographic arch separating the Appalachian and Michigan basins became emergent; populations of the species found in these basins became isolated on either side of that barrier, diverged, and then speciated, and a phylogeny relating all three species is shown overlying the basins. If sea level were to rise slightly such that the arch separating the Michigan and Appalachian basins became breached the new species in the Appalachian Basin could move into the Michigan Basin and vice versa.

turnover pulses should show congruent patterns of vicariance and geodispersal.

Now what about interpreting the differences between the vicariance and geodispersal trees? As discussed above, differences between them are good indication that noncyclical geological processes influenced biogeographic patterns. (Or, if these geological processes were cyclical, the time frame of the cycles would not be commensurate with the time frame we associate with speciation.) Noncyclical geological processes that can influence biogeographic patterns in such a manner as to produce differences between the vicariance and geodispersal trees are continental collisions. These would allow taxa to expand their ranges into new regions, but they would not allow for episodes of geodispersal followed by vicariance followed by subsequent geodispersal. However, divergences between the vicariance and geodispersal trees can also be produced by traditional long-distance dispersal. In this case one or more clades would have expanded its range over a geographic barrier as the result of some chance event, and there would be little or no prospect of returning to the original ancestral region.

The region whose position differs between the two Middle Devonian biogeographic trees is Kazakhstan, and here traditional dispersal is likely to be the best explanation, as Kazakhstan has never been implicated in the Acadian Orogeny and probably did not collide with either North America or Armorica at any time during Devonian period [see e.g., Scotese and McKerrow (1990) or Scotese (1997)].

From this example one can see that there are three fundamental pieces of information that can be extracted from such a biogeographic study: the role Earth history plays in motivating evolution, whether these events were cyclical or not, and the sequence of tectonic events. The results from another case study (Lieberman, 1997) enforce these principles, and are described below.

9.13.2. Early Cambrian Cladistic Biogeography and the Cambrian Radiation

The Cambrian radiation was one of the fundamental events in the history of life. Many of the metazoan phyla appear and diversify in the fossil record during this period; and the event appears to be closely associated with a time of profound metazoan cladogenesis at the highest taxonomic levels (Lieberman, 1999a). Coincident with these major biological changes is a series of profound tectonic events before and during the radiation, such that it is likely that the geological changes influenced the biological changes (Knoll, 1991, 1996; Signor and Lipps, 1992; Dalziel, 1997; Lieberman, 1997). The geological changes around this time include the breakup of the supercontinent Rodinia and the assembly of the supercontinent Gondwana. I investigated (Lieberman, 1997) biogeographic patterns in Early Cambrian trilobites using the analytical framework discussed herein to: (1) consider the relationship between the geological and the biological changes; (2) study the sequences of tectonic events during the breakup of Rodinia; and (3) determine whether cyclical geological events played a role in motivating biogeographic patterns.

In my biogeographic analysis (Lieberman, 1997) I focused primarily on the olenellid trilobites, a diverse group that appeared during the Cambrian radiation, and I subsequently presented phylogenetic hypotheses for a monophyletic clade within the olenellids (Lieberman, 1998, 1999b). This group was distributed throughout Laurentia (North America including Greenland), Siberia, and Baltica (Scandinavia) (Fig. 24). Around the time of the Cambrian radiation, these regions were rifting apart, which suggests that continental rifting was the primary geological process influencing biogeographic patterns in these trilobites.

The relative and absolute sequence of the rifting events separating Laurentia, Siberia, and Baltica have been worked out in general terms using paleomagnetic and geological data, but there still is controversy. In terms of the relative chronology of the tectonic events, some researchers have suggested that Siberia separated from Laurentia prior to the separation of Baltica (Scotese and McKerrow, 1990; McKerrow et al., 1992; and Torsvik et al., 1996), while others have maintained that Laurentia and Siberia were in contact after Laurentia and Baltica separated (Scotese, 1997). In terms of their absolute chronology, the separation of some or all of these three cratons has been put anywhere from the late Neoproterozoic (Scotese and McKerrow, 1990; Dalziel, 1991; Hoffman, 1991; McKerrow et al., 1992; Compston et al., 1995; Torsvik et al., 1996) to right at the Precambrian–Cambrian boundary (Hoffman, 1991; Condie and Rosen, 1994; Pelechaty, 1996).

The phylogeny used in the biogeographic analysis was based on my work (Lieberman, 1998, 1999b) but differs from the latter paper in that it includes

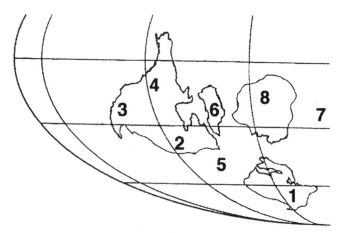

FIGURE 24. Paleogeographic map for the Early Cambrian. 1 = Baltica, 2 = eastern North America, 3 = southwestern North America, 4 = northwestern North America, 5 = northwest Britain, 6 = Greenland, 7 = northwestern Africa and 8 = Siberia. From Lieberman (1997), used by permission.

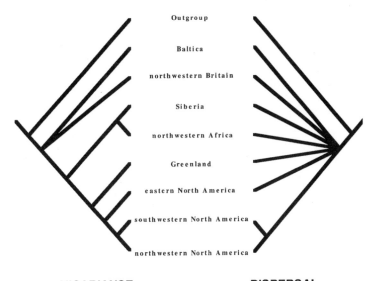

Outgroup

Baltica

northwestern Britain

Siberia

northwestern Africa

Greenland

eastern North America

southwestern North America

northwestern North America

VICARIANCE   **DISPERSAL**

FIGURE 25. Results from analysis of biogeographic patterns in Early Cambrian trilobites based on Lieberman (1999a). The vicariance tree is on the left and the geodispersal tree is on the right. There is little similarity between them and the latter is poorly resolved. From Lieberman (1999a), used by permission.

only two species of the genus *Olenellus*—*O. thompsoni* and *O. transitans*. At the time my first paper on the subject was published (Lieberman, 1997), the other species of the genus *Olenellus* had not yet been subjected to phylogenetic analysis. The phylogeny was converted to an area cladogram, which was converted to two BPA matrices as described above, and the data were analyzed using PAUP. The results are shown as a vicariance tree and a geodispersal tree, displayed side by side (Fig. 25). Various tests that I described (Lieberman, 1997) indicate that overall the different resolved aspects of the two trees are fairly well supported.

The vicariance tree can be used to tease apart the sequence of continental rifting in the Neoproterozoic and Early Cambrian. It first indicates that Laurentia shares a closer biogeographic relationship with Siberia and northwestern Africa than it does with Baltica, which suggests that the barrier separating Siberia from Laurentia emerged after that separating Baltica from both Siberia and Laurentia. Therefore, Laurentia and Siberia remained in contact after the separation of Baltica from these two regions, which supports the predictions of Pelechaty (1996) and Scotese (1997). However, when differences in the relative degree of support of different parts of the tree were quantified [see discussion in Lieberman (1997)], it turned out that the node joining Siberia and northwestern Africa to Laurentia is not a very strong one. This implies that Baltica may not have separated from Laurentia much before Siberia, so these results do not mediate too strongly against the hypotheses of Scotese and McKerrow (1990), McKerrow *et al.* (1992), and Torsvik *et al.*

(1996). Northwestern Britain was very likely connected to Laurentia in the Early Cambrian (Swett, 1981; Scotese and McKerrow, 1990; Fortey and Cocks, 1992). However, this region has very few species of olenelloid trilobites, so its position relative to the other regions is not well resolved (Lieberman, 1997). This is a common phenomenon in biogeographic studies: very low diversity regions tend to map to the basal part of the tree (Fortey and Cocks, 1992).

The results from the geodispersal tree suggest little congruent geodispersal: the tree is very poorly resolved. If we consider the mechanisms governing geodispersal and compare them with those governing vicariance we can see a clear link between Earth history events and patterns of evolution during the Cambrian radiation. From other lines of geological evidence we believe that during the Early Cambrian, Laurentia, Siberia, and Baltica were exclusively fragmenting from one another. Such continental rifting should overwhelmingly produce allopatric speciation and vicariance. By contrast, among the faunas of these regions there would have been few opportunities for geodispersal that might be associated with continental collision in the Early Cambrian. These signals are replicated faithfully in the biogeographic results—well-resolved patterns of vicariance and poorly resolved patterns of geodispersal—which is exactly what we would predict if Earth history and tectonic events were motivating patterns of evolution and distribution.

When comparing patterns from the vicariance and geodispersal trees, it is clear that there are very few similarities. The only commonality is the shared close relationship between northwestern and southwestern North America. Based on the precepts developed above, this indicates that the processes governing geodispersal were generally not the same as those that governed vicariance. Thus, in the Early Cambrian, cyclical geological and climatic processes had little influence on biogeographic patterns, except in the case of northwestern and southwestern North America. On the whole, repeated episodes of sea-level rise and fall played a limited role in influencing biogeographic patterns in the Early Cambrian. It has often been argued that sea-level rise played a fundamental role in driving the Cambrian radiation. These biogeographic data suggest that if this is true, the rise in sea level must have been so quick that it was biogeographically uninformative, and, further, there were not repeated episodes of sea-level rise and fall that later linked and then separated regions, with the exception of northwestern and southwestern North America.

9.13.3. Taking a Comparative Earth History Approach: the Difference between the Cambrian and the Devonian

Knoll *et al.* (1996) have argued that a comparative Earth history approach is an excellent way to study evolution in the fossil record. Indeed, such patterns form some of our fundamental data concerning the nature of the evolutionary process (Eldredge and Cracraft, 1980). A similar type of approach, when applied to biogeographic patterns, can also be very illuminating (Lieberman, 1999a). For example, biogeographic patterns in trilobites during the Middle Devonian were very different from those during the Early Cambrian, and the differences reflect different Earth history events. In the Early Cambrian, a vicariance signal dominated the biogeographic patterns (Fig. 25) because a supercontinent was being fragmented by rifting. In the Middle Devonian, geodispersal occurred frequently (Fig. 22) related to the repeated episodes of sea-level rise and fall that were occurring at the time.

From the biogeographic patterns it appears that there were two different types of Earth history regimes. One might also predict that the enhanced opportunities for vicariance and the limited opportunities for geodispersal in the Early Cambrian relative to the Middle Devonian might accentuate speciation rates in Early Cambrian taxa relative to their Devonian kin. This pattern was also found. Speciation rates in Early Cambrian trilobite taxa are significantly higher than those of Middle Devonian trilobites (Lieberman, 1999a). These elevated speciation rates are not enough, by themselves, to explain what was unique about the Cambrian radiation. However, they clearly would have contributed to this event, which is characterized by its explosive diversification in a geologically short period of time. Thus, just by considering the different Earth history regimes that prevailed at different times, each of which has a distinct biological and biogeographic signature, one can gain significant insight into major events in the history of life. The importance of differences in Earth history events in shaping biogeographic patterns indicates that these events have a fundamental role in shaping patterns of biological evolution. This demonstrates that study of the fossil record can make important contributions to our understanding of the evolutionary process.

9.14. Further Issues in Cladistic Biogeography that Need to be Explored

One of the fundamental aims of any biogeographic study is to ascertain the role that Earth history plays in influencing the evolutionary process, and there are many types of geological and climatic events that can influence biogeographic patterns. Different events can produce very divergent patterns, e.g., vicariance and geodispersal caused by the fragmentation and collision of plates, respectively. One set of processes may prevail for a long time, but then another set, which might be expeceted to engender very different biogeographic patterns, may come to prevail. More critical for biogeographic studies is the fact that not only the processes may change, but how they influence the interactions among different regions may change as well. For example, if we look again at the case of the different tectonic basins in ENA, we see that repeated episodes of sea-level rise and fall can unite and then isolate the basins in the same sequence again and again over many millions of years. However, we can now imagine that a tectonic change related to the Acadian Orogeny occurred, which changed the geometry of the arches that separate the basins from one another such that the barrier between the Appalachian Basin and the others became more prominent, whereas that between the Illinois and Michigan basins became less prominent. The patterns of vicariance and geodispersal that ensued before the tectonic changes are those shown in Fig. 23. The biogeographic patterns that would ensue after the change would be different. Now the Illinois and Michigan basins would group with one another to the exclusion of the Appalachian on both the vicariance and the geodispersal trees. If we analyze biogeographic patterns in groups that persist through several of these different types of Earth history regimes we might expect to lose some of the biogeographic signal in the data. There might be conflicting signal, with some evidence for a closer relationship between the Appalachian and Michigan basins, and some evidence for a closer relationship between the Michigan and Illinois basins. These divergent events might not efface all the biogeographic signal in the data and might not cause a complete loss of resolution in the geodispersal and vicariance trees, but they would certainly reduce the amount of support for the most-parsimonious trees. Is this a problem that should be remedied by biogeographic studies, and if so, how can it be?

Clearly, this problem will become most acute in the study of clades that persist through long periods of time and are thus more likely to be governed by very different tectonic scenarios. For example, a study that attempts to resolve biogeographic patterns in living lungfish may not be well advised, but can we break up the analysis of long-lived groups into several different subclades in discrete time periods bracketed by different episodes of tectonic change? Part of the difficulty with this is that we are often using the biogeographic patterns to investigate the tectonic events in the first place, and we do not necessarily know the nature of the tectonic events or whether one time period is characterized by one type of event and another time period by another. Therefore, this might be considered ill-advised. Further, to divide a large clade up into separate subclades might lead to the creation of paraphyletic groups, a violation of cladistic principles. Unless there was a clear case for recognizing some of the taxa in one time period as the immediate ancestors of taxa in the succeeding time period, this could lead to spurious results, because some of the taxa that should be considered in the biogeographic analysis of each time period would actually be excluded. Thus, this solution to the problem is not generally recommended.

Fortunately this problem is often obviated by extinction, which is one of the fundamental phenomena in the history of life, first recognized from study of the fossil record: most species and clades have very finite durations. They tend to go extinct, and often the cause of the extinction is a major episode of environmental change, which can be partly driven by tectonic changes, thus giving rise to a natural sorting. This is exactly the phenomenon of coordinated stasis described in Chapter 2. If coordinated stasis operates as a general principle organizing the nature of the fossil record, then it bodes well for our ability to constrain biogeographic problems to tractable intervals of time. Different tectonic regimes will tend, if this is true, to bracket different clades, and well-resolved biogeographic patterns produced by congruent geological processes will fall out as a natural result.

Another way of surmounting this problem is to only consider clades whose duration is limited in biogeographic studies. This can be achieved by choosing clades of the appropriate taxonomic level. This solution is in a sense analogous to that of the problem encountered in molecular systematics: when analyzing the evolutionary relationships of a group of organisms it is necessary to choose the right gene or genes. If a gene that evolves too rapidly is chosen, one may get poor resolution in the molecular gene tree. If a gene that evolves too slowly is chosen, no differences, or not enough differences, will emerge among the various component species being considered, and the gene tree will also be poorly resolved.

If this seems vaguely unsatisfactory, perhaps that is because it is. Different taxonomic categories above the species level have little or no meaning and they are often created by simple fiat. However, for now this may be the best way to proceed, especially recognizing that extinction in the fossil record also does much of the necessary sorting beforehand, parsing clades into distinct time periods marked by significant environmental change. In any event, this is obviously a topic that needs much more attention from both a theoretical and an analytical perspective. For now, how we should divide up biogeographic analyses into geologically cohesive time periods remains just as ill-defined and just as crucial a topic as the problem of defining areas that was discussed in Chapter 7.

9.15. Conclusions

The problem of defining a specific time period in the fossil record for a paleobiogeographic study is an important one, but it would be wrong to end this chapter on a negative note. Paleobiogeography emerges as a discipline that has many strengths compared to its neontological cousin. In particular, rampant extinction, which can leave the modern biota highly pruned, greatly influences biogeographic patterns recovered from extant organisms and can lead to erroneous results if the investigator is not aware of the extinct taxa. By contrast, in the fossil record, these extinct taxa are generally all that is observed. In addition, the opportunity for unprecedented synthesis exists in the areas of paleobiogeography and biogeography. This is largely due to the many theoretical and analytical advances in biogeography, which now allow researchers to consider a range of problems in geology and evolutionary biology. Chief among them are the documentation of sequences of geological events, the assessment of the presence or absence of cyclical climatic or geological phenomena, and the demonstration of a close link between Earth history events and evolution. Studies that have been conducted thus far have demonstrated that there is clearly a potential for retrieving resolved biogeographic patterns. Moreover, there are opportunities for illuminating interesting biological and geological questions by comparing biogeographic patterns at different times in Earth history.

However, in order to be successful, biogeographic methods must provide a way to take congruent episodes of both vicariance and geodispersal into account. A method that fails to consider geodispersal, e.g., components analysis, will be inadequate, as it will generally not retrive any resolution in a biogeographic study because a pure signal of vicariance, if not analyzed correctly, will be swamped by geodispersal. Even when some resolution is achieved by studies that use components analysis, it will generally be artefactual, except in the rare cases when the studies are conducted on regions where geodispersal has never occurred.

Part of the reason researchers have been disinclined to conduct biogeographic studies is that they have been discouraged by the inability of some analytical methods to retrieve resolved biogeographic patterns, a pessimism that followed closely on the heels of the initial enthusiasm associated with the development of cladistic biogeography. However, if we use a method of biogeographic analysis that takes geodispersal into account, and if we focus on the fossil record in order to avoid the problems that extinct taxa can cause, this pessimism should evaporate. In short, not only does the opportunity for unprecedented synthesis exist, but there is also an opportunity for a disciplinary renaissance.

Chapter 10

Biogeography and the Biodiversity Crisis

10.1. Introduction

Clearly one of the greatest challenges facing humanity in the next century is the biodiversity crisis and the associated degradation of the Earth's biosphere. Biogeography can contribute to our understanding of this crisis in several important ways because the mechanisms governing the current mass extinction are partly related to biogeographic phenomena. For example, among the primary causes of the precipitous loss of biological diversity are invasive species (Brown and Lomolino, 1998; Cohen and Carlton, 1998; Eldredge, 1998), i.e., species that are moved by humans, accidentally or intentionally, to new regions. When these new species arrive in endemic areas they sometimes proliferate to the detriment of the taxa already present. There are numerous examples of this phenomenon [see, e.g., Elton (1958) and Cohen and Carlton (1998)], and invasive species are estimated to cost $123 billion a year through their negative effects on crops, human health, and lost commerce (Stolzenburg, 1999). In fact, the problem has become so serious that in February 1999, President Clinton signed an executive order that established the Invasive Species Council to counter it.

One particularly well known example of an invasive species is the zebra mussel, which colonized the rivers and lakes of North America from Europe some time in the 1980s. North America harbors one of the most diverse freshwater molluscan faunas in the world, including the unionid bivalves, but the arrival of the zebra mussel has caused significant problems for the unionids. It has proliferated to such an extent that in certain regions it has

177

completely overgrown everything else in aquatic habitats, including the unionids. Not only does this smother the unionids, but the tremendous densities that the zebra mussels reach change water composition and chemistry, fundamentally altering aquatic ecosystems.

Vermeij (1978) documented several examples of how invasions in marine invertebrate biotas related to the construction of the Suez and Panama canals had extreme consequences for biodiversity. There have also been detailed studies conducted on the San Francisco Bay and Delta that show the phenomenon of invasive species in alarming detail. Cohen and Carlton (1998) reported the presence of at least 234 exotic species in this large estuary, some of which comprise 99% of the local biomass. Further, the rates of invasion in this estuary are truly astounding. Between 1851 and 1960 a new species became established roughly once every year; after 1961 the rate increased to a new species roughly once every 3 months (Cohen and Carlton, 1998), perhaps related to the increasing globalization of the economy.

Brown and Lomolino (1998), based on the results of Flather *et al.* (1994), analyzed the basic factors causing historical animal extinctions; they found that 38% could be attributed to accidentally or purposefully introduced species and that this was the single largest causal factor. However, they recognized that habitat destruction has begun to overtake invasive species as the primary factor causing extinction, and the role that it has played in precipitating the current biodiversity crisis has been thoroughly documented, [see, e.g., the accounts in Wilson, (1988, 1993, 1994), Eldredge (1991, 1992, 1998), or in any issue of *Nature Conservancy* magazine].

Both invasive species and habitat destruction can be viewed as types of biogeographic phenomena which will be discussed more fully shortly. In fact, simply from our understanding of biogeographic principles we would predict that these factors would engender wholesale extinction. Moreover, apart from all the studies showing how invasive species and habitat destruction affect the modern biota, there is evidence that they also had an impact at other times in Earth history.

10.2. Invasive Species and the Biodiversity Crisis: Geodispersal and Merging Areas of Endemism

In biogeographic terms we can treat invasive species as effectively homogenizing areas of endemism, as their geographic ranges are being extended via a phenomenon akin to geodispersal. In this case, it is our own species that is effectively removing the geographic barriers. For example, the barriers that once checked the free movement of many taxa are being erased through global commerce and travel, with the result that the total number of areas of endemism on Earth is decreasing. Naturally, we would predict that this would lead to a concomitant decline in global diversity. There are several paleobiogeographic phenomena in the history of life that bear evidence of this.

We can think back to the correlation between plate tectonic events and biological diversity: separate continents, amplify the number of independent regions, and total diversity goes up; do the reverse and diversity decreases. There is also the distinction between paleobiogeographic patterns in the Cambrian and in the Devonian, which indicated that there was more geodispersal in the Devonian than in the Cambrian, and thus speciation rates were lower in the Devonian. Thus, even if we do not include the many other negative impacts human beings are having on the Earth's biota, the simple act of accelerating geodispersal and decreasing the number of areas of endemism would lower global diversity and precipitate a biodiversity crisis. However, the additional negative impacts such as pollution and general habitat degradation are critical in this regard. Because the chances of different regions being invaded by nonnative species, and the likelihood of these invasions causing extinctions, greatly increase when the indigenous biotas are disturbed (Brown and Lomolino, 1998; Cohen and Carlton, 1998). We are not only facilitating geodispersal, but are making it more likely that it will lead to extinction. Further, it is the regions with the most endemic faunas, often island biotas, that show the highest proportion of biological invasions.

10.3. Analogues from the Past: the Late Devonian Mass Extinction

If we look back through the history of life we can see that there are times of mass extinction that correspond paleobiogeographically to the present-day biodiversity crisis. For example, there is at least one other mass extinction that may have been partly caused by something like the invasive species phenomenon. The Late Devonian mass extinction, one of the five greatest such events in the history of life (McGhee, 1988, 1989), was also a time when many marine invertebrate taxa were cosmopolitan. Their geographic ranges had increased in the Late Devonian relative to the Middle Devonian (Oliver, 1977, 1990; Oliver and Pedder, 1989, McGhee, 1996) such that there were far fewer areas of endemism. From simple biogeographic principles we would predict that this would reduce overall biological diversity, and at this time there was a great drop in the diversity of marine invertebrates, which was sufficient to have produced a mass extinction.

The decline in the number of areas of endemism in the Late Devonian can be explained by Earth history events. There were several episodes of profound sea-level rise at this time (McGhee, 1988, 1989, 1996; Hallam, 1992), and continuing continental collisions related to the Acadian Orogeny. As described above, these would have increased opportunities for geodispersal in marine invertebrate taxa, and that would reduce the number of areas of endemism and thus diminish biological diversity. This is not to say that there were no other changes occurring in the Late Devonian that could have contributed to the mass extinction. There appear to have been several major environmental perturbations throughout the period [McGhee (1988, 1989, 1996); also see Kirchgasser *et al.* (1985)], which may have also played important roles in mediating the mass extinction. However, it is clear that some of the governing mechanisms are related to the decline in the number of areas of endemism.

Now what does this mean in terms of the current biodiversity crisis? In a sense, the biogeographic impact of our own species is as profound as the biogeographic impact of major Earth history events felt at other pivotal times in the history of life. The simple presence of human-induced geodispersal alone could act to produce mass extinction, everything else being equal.

10.4. Habitat Destruction and the Biodiversity Crisis: Destroying Areas of Endemism

Of course everything else is not equal in the biodiversity crisis. Not only are we homogenizing areas of endemism, but we are eliminating them entirely through habitat destruction. These two biogeographic phenomena act in concert to have an even greater role in causing extinction. If a large enough region is paved over and converted to suburban housing or a large enough area of tropical rain forest is burned, then an entire area of endemism with all of its biological diversity must be eliminated. For example, in the northeastern United States there is a single great urban corridor running from Portland, Maine, to Washington, D.C. This is an enormous swath of land encompassing thousands of square miles that is largely biologically depauperate. A few hundred years ago it would have been largely pristine wilderness.

Just as the role of invasive species in the current biodiversity crisis is illuminated by surveying past biodiversity crises, so too is habitat destruction. For example, the cause of the mass extinction at the Cretaceous–Tertiary boundary was a large asteroid striking the planet (Raup, 1986), which was habitat destruction on a global scale. The mechanism driving the Permo–Triassic mass extinction may have been upwelling carbon dioxide from the ocean bottom to the continental shelf, which poisoned diverse ocean biotas (Knoll *et al.*, 1996), which was again, habitat destruction on a global scale. Global cooling played an important role in precipitating the Ordovician–Silurian mass extinction (Stanley, 1987), as it greatly constricted the area that could be occupied by diverse tropical biotas, while degrading their preferred habitat.

10.5. Historical Perspective on Biogeography's Role in Understanding the Biodiversity Crisis

Early major figures in the field of biogeography recognized the impact of human beings as exponents of biogeographic change and the profoundly negative impacts that these changes have had on the biosphere. Lyell (1832), who was one of the most insightful early thinkers in this connection, realized that by moving species around the globe, human beings were altering the balance of nature. "Man is, in truth, continually striving to diminish the natural diversity of the stations of animals and plants in every country, and to reduce them all to a small number fitted for species of economical use" (Lyell, 1832, p. 147–148). He also described how humans were currently extirpating many local populations of organisms and how they had already in his day driven other species to extinction. "We must at once be convinced, that the annihilation of a multitude of species has already been effected, and will continue to go on hereafter, in certain regions, in a still more rapid ratio, as the colonies of highly civilized nations spread themselves over unoccupied lands" (Lyell, 1832, p. 156).

It is a sobering thought that humanity's role in the biosphere was recognized as early as the 19th century, albeit by one of the great intellectuals of his day. Lyell's (1832) predictions have come true, and the dire effects of our activities continue to accelerate. Even if our own species continues to prevail in the face of this massive-scale destruction, what will be the state of the world we live in? Van Oosterzee (1997) poignantly captured this idea when she revisited some of the sites in the Malay Archipelago, the once richly diverse tropical regions that inspired Alfred Russell Wallace, the father of evolutionary biogeography. It was his travels and collecting activities in this region that were so crucial in his evolutionary and biogeographic discoveries. Van Oosterzee (1997) saw that the biotas that had been so pivotal for Wallace were now almost completely destroyed as a result of human exploitation and development. The scientific process is a creative one, so where will the great evolutionary biologists and ecologists of the future turn for their sources of inspiration? If we lose this creative spark for intellectual activity we lose a part of our humanity, not to mention a rich source of beauty and wonder, as well as a repository of many of the Earth's medicines (Wilson, 1994; Van Oosterzee, 1997). Yet it almost seems like folly to sound a warning call, recognizing that Lyell's (1832) observations, made so long ago, went completely unheeded.

10.6. Biogeography: More Lessons from the Recent Past

As discussed in Chapter 4, one of the fundamental revolutions in our understanding of the Earth's biota that contributed to the birth of biogeography as a scientific discipline was the discovery that very different types of organisms existed in different regions. This was discovered during an era of colonial investigation and exploration. Perhaps the greatest of the expeditions, in terms of its impact on science and natural history, was on the H.M.S. *Endeavor*, the boat captained by James Cook, between 1768 and 1771. The natural historian on board was Joseph A. Banks, later the *de facto* director of the Royal Botanic Gardens at Kew and the president of the Royal Society of

London. Banks visited many places, including Brazil, Tahiti, New Zealand, Australia, New Guinea, Indonesia, and South Africa. The biotas were biologically diverse regions, areas of endemism that harbored many distinct taxa, and he collected everything unique. Specimens discovered during this voyage increased the number of plant species known at the time by almost 25 percent (Watkins, 1996). This truly astounding percentage is equivalent to over 1400 species; perhaps 2200 new animal species were discovered as well.

However, not only did Banks expand our knowledge of the biotas of these areas of endemism. Ironically, he initiated a campaign of moving species from one area of endemism to another. For example, in his journal he noted how he planted seeds of plants from Brazil in Tahiti (Watkins, 1996). He was thus in the very act of breaking down the biological richness that he catalogued. In the course of its discovery, Banks was effacing the very reason that this diversity existed. Effectively, he was undertaking large-scale geodispersal, artificially erasing the geographic barriers that had allowed biotas to evolve in isolation, which had led via vicariance to their diversity. Even more ironic, some of the leaves of the new plant taxa he discovered were pressed between sheets of a copy of a commentary on Milton's *Paradise Lost* (Watkins, 1996). This is not intended as condemnation: Banks was a man of his time, but in more ways than one paradise was both discovered and thus ultimately lost after his voyage with Cooke.

More than any other man, Banks can be credited with initiating the homogenization of areas of endemism on a global scale. He was, e.g., one of the chief backers of the scheme to send convicts to Australia; along with these convicts were shipped cattle, sheep, and many species of agriculturally relevant plants (Watkins, 1996). This initiated a policy of biological invasion into the Australian continent on a grand scale, and the Australian biota, with its many diverse areas of endemism, will never recover from it. Implicit in Banks' important discoveries was the birth of biogeography and also the seeds of its destruction. Areas of endemism, the very phenomenon that would later demonstrate the importance of geography as a factor influencing the evolution of life, were recognized and slowly, yet inexorably, destroyed.

10.7. Conclusions

The principles of biogeography point the way to a basic theoretical understanding of the biodiversity crisis. However, while these principles suggest an easy solution, it will be difficult or impossible to implement it. Whether humanity will have the moral resolve to not just say stop but actually do it will determine the future of biodiversity and also our own fate. Will we have a biological heritage that future generations can cherish, or will we leave them with a biologically impoverished world of little inspiration?

Pivotal figures in biogeography have long recognized the important role of areas of endemism as centers of biological diversity. More disturbingly, some of these figures even worked assiduously to initiate processes that disrupted these centers. In the end, be it either by homogenization or the actual destruction of areas of endemism, we have embarked on a great experiment with the Earth's biota, but it is an experiment that has been conducted before. The lessons from the past indicate several major periods of mass extinction whose mechanisms are coupled in various ways to the destruction of the biogeographic centers that generate diversity. The lesson from the fossil record suggests that there was eventual recovery, in terms of total species diversity, though not of actual taxa, but the timescale was tens of millions of years.

It is a trite truism, but those who do not learn the lessons of history are doomed to repeat them. The impressive lag time between the destruction and recovery of global ecosystems suggests that our species will be paying the price of this biodiversity crisis for many generations to come, even if we survive this sixth great mass extinction. In the immense span of geological time our own species' impact, and that of every individual to a much lesser extent, will be as ephemeral as the moisture that one's breath leaves on a window in winter after familiar surroundings are fondly surveyed. But for our children and our children's children these impacts will be real and irreversible on the timescales of experience. The time has come to reverse the spreading seeds of destruction that our culture has initiated. If our surroundings are to be worth surveying for inspiration in the future, we have to rebuild the walls that have maintained and generated biological diversity in the past and improve and enhance the conditions behind those walls.

Chapter 11

Conclusions

Biogeography and paleobiogeography are disciplines that have made, and will continue to make, important contributions to biology and geology. The uniquely interdisciplinary roles of biogeography and paleobiogeography mean that a practitioner in these areas should be able to converse meaningfully with ecologists, evolutionary biologists, and geophysicists. This creates special challenges but also special opportunities. Biogeography, and, by extension, paleobiogeography, has traditionally been divided into two subdisciplines—ecological biogeography and historical biogeography—and the distinction between them has a long intellectual history. Traditionally, these two fields were so defined that ecological biogeographers concern themselves with how ecology influences biogeographic patterns and historical biogeographers study how geological and climatic changes influence biogeographic patterns. These definitions have tended to create barriers between the two subdisciplines.

When thinking about the diversity of interests encompassed in biogeography it is helpful to consider the fact that life is made up of two distinct sets of hierarchically arrayed entities. Ecological and historical biogeography should be redefined to coincide with the study of biogeographic patterns in entities of the economic and genealogical hierarchies. This approach shows when ecological and historical biogeographic patterns diverge and when they are equivalent. The common entities the two hierarchies share serve as a way to unite these long-separated subdisciplines, and processes such as coevolution are natural links between them. Further, the study of biogeographic patterns in each entity will encompass a specific spatial and temporal scale, each temporal scale representing a mode in which climatic or geological cycles operate.

This hierarchy can also be used to help define paleobiogeography as a subdiscipline within biogeography. Of course, paleobiogeography is just the study of biogeographic patterns in the fossil record, but it is well known that the fossil record is incomplete. Thus we can only study biogeographic patterns in certain types of genealogical and economic entities: the larger-scale entities such as clades and regional biotas and possibly species and communities. Paleobiogeographic patterns cannot be studied in small-scale entities of either hierarchy because their temporal duration is shorter than our limits of resolution in the fossil record. Thus, the distinction between biogeography and paleobiogeography is only one of hierarchical and temporal scale.

The types of differences that emerge between ecological and historical paleobiogeography have to do with whether or not economic and genealogical entities at the higher hierarchical levels diverge in their patterns of stability and persistence over long timescales. It has already been thoroughly established that large-scale genealogical entities persist over long timescales, but whether the same is true of economic entities has to be determined. For example, if large-scale economic entities do not persist through time, it will be impossible to study paleobiogeographic patterns in them.

The debate among paleontologists about the existence of coordinated stasis is relevant to this issue. Although coordinated stasis has not always been framed as a problem with paleobiogeographic relevance, the debate about it is partly about the biogeographic coherence of large-scale ecological or economic entities such as regional biotas. If it is valid, then large-scale economic entities will persist and show paleobiogeographic patterns over long time spans.

The fundamental pattern that historical paleobiogeographers are looking to uncover is congruence: similar patterns of changing geographic distribution associated with evolutionary change in different groups. These patterns are expressed as area cladograms, where a species from one area is the closest relative of a species from a different area, and in different groups they are likely to reflect large-scale geological or climatic changes that have influenced different groups of organisms in a similar way.

Originally, it was believed that biogeographic patterns in different groups of organisms could only result from vicariance. However, in reality, there are two fundamental processes that can produce congruent paleobiogeographic patterns, for in addition to vicariance there is geodispersal. In vicariance congruence results from the formation of geographic barriers, while in geodispersal it is produced when geographic barriers fall.

The recognition that geodispersal can lead to biogeographic congruence actually provides another link between ecological and historical biogeography because traditionally it had been argued that range expansion and dispersal were processes relevant only to ecological biogeographers. However, if range expansion can produce biogeographic congruence then it should also be of significance to historical biogeographers. Then range expansion becomes a process that runs across both hierarchies and thus has relevance to both ecological and historical biogeography. If historical biogeographers do not acknowledge both types of processes, then it will be very hard to obtain meaningful results in a biogeographic study.

The importance of geodispersal as a biogeographic process also affects the way we define areas in biogeographic studies. Areas have traditionally been analogized to clades because it was thought that biogeographic patterns could only be produced by vicariance and thus occurred via a phenomenon akin to evolution via common descent. In this view, just as a clade differentiates as it evolves, in biogeography a larger area would split into smaller areas owing to the emergence of geographic barriers. The most recently separated areas would share the most recent common geological and climatic history, just as the taxa that shared a most recent common ancestor shared a most recent common evolutionary history. This is clearly a common biogeographic mode. However, because of geodispersal, areas also agglomerate into larger areas. Thus, the history of areas is not solely equivalent to strict common descent or phylogeny. Instead, they show something akin to what in systematics is called tokogeny. Tokogenetic relationships occur when taxa can interbreed, and this hybridization merges what were once separate lineages. One type of taxon in systematics that shows both phylogenetic and tokogenetic patterns of relationship is species. Areas are akin to species; their parts can diverge, but they can also converge.

Because the species problem has been perhaps the most vexing problem in the history of biology, we should recognize that the problem of defining areas will encounter similar difficulties. However, one crucial aspect of the debate about the nature of species is that we have to consider both their ontology and their epistemology. This also holds true if we want to truly understand and solve the problem of defining areas. First, what is the nature or ontology of areas; what gives them their existence? Second, how do we recognize areas; how should we treat them epistemologically? The ontology of areas derives from the fact that the presence of geographic barriers encourages evolutionary change, and the absence of barriers allows taxa to expand their ranges. Areas have an existence related to their independent geological history, which will concomitantly impact patterns of biological evolution. Epistemologically, we can recognize areas in several ways. They might circumscribe the geographic ranges of many taxa; they might be surrounded by prominent geographic barriers; they might contain a unique set of geological features that makes one believe that they were once geologically independent.

Biogeography is fundamentally linked to systematic biology. The fact that areas are akin to species, and can show a pattern of evolution, demonstrates one important part of this link. Another important part is that the best techniques available for use in biogeographic analysis are derived from systematics. In biogeography, as in systematics, it is of quintessential importance that rigorous analytical techniques be used. The absence of these techniques in the field of systematics, until fairly recently, was a serious impediment to any scientific progress, and the same was true of biogeography. Without analytical techniques, arguments about biogeographic patterns deteriorate into a series of "just-so" stories. Of all the techniques available BPA is the only method that can encompass the phenomenon of geodispersal. Moreover, it approaches the analysis of biogeographic patterns in the most legitimate and assumption-free manner. The geographic distribution of taxa and how that distribution changes as a group evolves are the data that BPA uses to interpret biogeographic patterns. In this book, I proposed that a few modifications be made to BPA as traditionally implemented, particularly that it be altered so that it can retrieve episodes of geodispersal. However, its basic framework remains intact.

When we consider the assumptions needed to conduct a biogeographic study, some of the distinctions between paleobiogeography and biogeography emerge, even though I have argued that these two disciplines are inextricably linked. It is clear that any study of the modern biota has significant strengths relative to studies of the fossil biota. However, an important assumption of a biogeographic study that considers only extant taxa is that extinction is minimal, because extinct taxa cannot be sampled. Any group that persists over geological timescales must have experienced some extinction, and this will artefactually influence perceived biogeographic patterns among the remaining extant members of that clade. Of course, the only way we can consider extinct taxa is by recourse to the fossil record. Thus, any biogeographic study that considers only extant taxa is potentially flawed.

Another potential flaw in historical biogeographic studies that focus only on extant taxa is that they cannot consider how patterns of area relationship might have changed through time. For instance, from the Cambrian to the Recent the North American continent experienced several cycles of continental rifting and collision. If we were to study biogeographic patterns in the extant biota of North America, which was last influenced by the most recent tectonic events there, we might see biogeographic patterns related to these events; however, in clades that have persisted through several major tectonic events in North America, there may be biogeographic patterns that are also related to earlier events. If these patterns are different from those caused by the more recent events, there will be pervasive biogeographic noise.

The strength of paleobiogeography is that we can actually trace paleobiogeographic patterns through time in North American fossil organisms. This allows us to get a much more complete picture of how all the tectonic events that occurred in North America from the Cambrian to the Recent have influenced the biogeographic patterns. We would also expect to find far less noise in paleobiogeographic studies than in biogeographic studies of the extant biota, because in the former we can concentrate on one time period and one tectonic event at a time. Thus there would be less concern that different tectonic events, which occurred at different times, had influenced the patterns.

Seen from this perspective, paleobiogeography emerges as a discipline of great strength. It has a long, rich tradition of making important contributions to the field of geology by increasing our understanding of tectonic events. In addition, biogeography was one of the disciplines that contributed greatly to the formulation of early ideas on evolution, and when paleobiogeography is integrated into this tradition it has the potential to make important contributions to evolutionary biology. Specific contributions would include documenting the role that Earth history events play in shaping large-scale patterns of evolution and in influencing rates of speciation.

Finally, paleobiogeography can help us understand the biodiversity crisis. Clearly the human species is responsible for this crisis, and our activities have contributed to extinction in many different ways. First, we are causing habitat destruction, equivalent to the elimination of areas of endemism. Second, either intentionally or accidentally, we are transporting animal and plant species around the globe, which is highly detrimental, and can be thought of as the homogenization of areas of endemism. Throughout Earth history the number of different areas of endemism correlates well with global diversity. By reducing the number of available areas of endemism in the modern biota, we are reducing global diversity.

In the history of life there have been five major mass extinctions, and in all probability the causes of each were related in some way to habitat degradation. However, there is evidence that homogenization of areas of endemism played a role in at least one of the five. This one case, the extinction in the Late Devonian, shows a signature similar to the one in the current biodiversity crisis, where a reduction in the number of areas of endemism owing to homogenization produced a concomitant decline in diversity. Insight into this mass extinction as a paleobiogeographic phenomenon may provide an understanding of the biogeographic mechanisms of the modern biodiversity crisis, particularly those relating to invasive species.

Biogeography emerges as a discipline that had its inception in the age of discovery. Early natural historians recognized that new lands harbored new species. Ultimately, the rich biological diversity of distinct regions, which was partly generated by biogeographic processes, proved inspirational to many natural historians, both personally and scientifically. Creativity is as much an element of science as art. The greatest, deepest insights in science and the noblest reflections of the human spirit involve creative thinking at its zenith. The study of biogeography reached that zenith in the work of a few of the greatest scientists in the West—Lyell, Wallace, and Darwin. For them it was a key that could unlock the history of the Earth and the history of life, a

> study which presents problems as vast, as intricate, and as interesting as any to which the human mind can be directed, where objects are as infinite as the stars of heaven and infinitely diversified, and whose field of research extends over the whole earth, not only as it now exists, but also during the countless changes it has undergone from the earliest geological epochs. (Wallace, 1857, p. 159).

References

Agassiz, L., 1840, *Études sur les glaciers*, Neuchatel.

Agassiz, L., 1842, On the succession and development of organic beings, *Edinburgh N. Phil. J.* **33**:388–399.

Allen, T. H. F., and Starr, T. B., 1982, *Hierarchy: Perspectives for Ecological Complexity*, University of Chicago Press, Chicago.

Allmon, W. D., 1992, A causal analysis of stages in allopatric speciation, *Oxford Surv. Evol. Biol.* **8**:221–257.

Anders, M. H., Krueger, S. W., and Sadler, P. M., 1987, A new look at sedimentation rates and the completeness of the stratigraphic record, *J. Geol.* **95**:1–14.

Arnold, A. J., and Fristrup, K., 1982, The theory of evolution by natural selection: a hierarchical expansion, *Paleobiology* **8**:113–129.

Avise, J. C., 1986, Mitochondrial DNA and the evolutionary genetics of higher animals, *Phil. Trans. Roy. Soc. London, Ser. B* **312**:325–342.

Avise, J. C., 1992, Molecular population structure and the biogeographic history of a regional fauna: a case history with lessons for conservation biology, *Oikos* **63**:62–76.

Avise, J. C., Arnold, J., Ball, R. M., Bermingham, E., Lamb, T., Neigel, J. E., Reeb, C. A., and Saunders, N. C., 1987, Intraspecific phylogeography: the mitochondrial DNA bridge between population genetics and systematics, *Ann. Revs. Ecol. Syst.* **18**:489–522.

Axelius, B., 1991, Areas of distribution and areas of endemism, *Cladistics* **7**:197–199.

Babcock, L. E., 1994, Biogeography and biofacies patterns of Middle Cambrian polymeroid trilobites from North Greenland: palaeogeographic and palaeo-oceanographic implications, *Bull. Grønl. geol. Unders.* **169**:129–147.

Barrett, P. H., Gautrey, P. J., Herbert, S., Kohn, D., and Smith, S., 1987, *Charles Darwin's Notebooks, 1836–1844*, Cornell University Press, Ithaca.

Barrett, M., Donoghue, M. J., and Sober, E., 1991, Against consensus, *Syst. Zool.* **40**:486–493.

Barrett, M., Donoghue, M. J., and Sober, E., 1993, Crusade? A reply to Nelson, *Syst. Biol.* **42**:216–217.

Bayer, U., and McGhee, G. R., Jr., 1988, Evolution in marginal epicontinental basins: the role of phylogenetic and ecological factors, in: *Sedimentary and Evolutionary Cycles* (U. Bayer and A. Seilacher, eds.), Springer-Verlag, New York, pp. 164–220.

Beaumont, C., Quinlan, G., and Hamilton, J., 1988, Orogeny and stratigraphy: numerical models of the Paleozoic in the eastern interior of North America, *Tectonics* **7**:389–416.

Beddard, F. E., 1895, *A Textbook of Zoogeography*, Cambridge University Press, Cambridge.

Bennett, K. D., 1990, Milankovitch cycles and their effects on species in ecological and evolutionary time, *Paleobiology* **16**:11–21.

Bennington, J. B., and Bambach, R. K., 1996, Statistical testing for paleocommunity recurrence: Are similar fossil assemblages ever the same?, *Palaeogeog. Palaeoclim. Palaeoecol.* **127**:107–134.

Benton, M. J., and Simms, M. J., 1995, Testing the marine and continental fossil records, *Geology* **23**:601–604.

Benton, M. J., and Storrs, G. W., 1994, Testing the quality of the fossil record: paleontological knowledge is improving, *Geology* **22**:111–114.

Berger, A., 1980, The Milankovitch astronomical theory of paleoclimates: a modern review, in: *Vistas on Astronomy* (A. Beer, K. Pounds and P. Beer, eds.), Pergamon Press, London, pp. 103–122.

Boer, A. J. de., and Duffels, J. P., 1996, Historical biogeography of the cicadas of Wallacea, New Guinea and the West Pacific: a geotectonic explanation, *Palaeogeog. Palaeoclim. Palaeoecol.* **124**:153–177.

Bond, G. P., Nickeson, P. A., and Kominz, M. A., 1984, Breakup of a supercontinent between 625Ma and 555Ma: new evidence and implications for continental history, *Earth Plan. Sci. Let.* **70**:325–345.

Bowler, P. J., 1996, *Life's Splendid Drama*, University of Chicago Press, Chicago.

Bremer, K., 1992, Ancestral areas: a cladistic reinterpretation of the center of origin concept, *Syst. Biol.* **41**:436–445.

Bremer, K., 1994, Branch support and tree stability, *Cladistics* **10**:295–304.

Bremer, K., 1995, Ancestral areas: optimization and probability, *Syst. Biol.* **44**:255–259.

Brett, C. E., and Baird, G. C., 1995, Coordinated stasis and evolutionary ecology of Silurian to Middle Devonian faunas in the Appalachian Basin, in: *New Approaches to Speciation in the Fossil Record* (D. H. Erwin and R. L. Anstey, eds.), Columbia University Press, New York.

Brooks, D. R., 1981, Hennig's parasitological method: a proposed solution, *Syst. Zool.* **30**:229–249.

Brooks, D. R., 1985, Historical ecology: a new approach to studying the evolution of ecological associations, *Ann. Miss. Bot. Gard.* **72**:660–680.

Brooks, D. R., 1988, Scaling effects in historical biogeography: a new view of space, time, and form, *Syst. Zool.* **37**:237–244.

Brooks, D. R., 1990, Parsimony analysis in historical biogeography and coevolution: methodological and theoretical update, *Syst. Zool.* **39**:14–30.

Brooks, D. R., and McLennan, D. A., 1991, *Phylogeny, Ecology, and Behavior*, University of Chicago Press, Chicago.

Brooks, D. R., and Wiley, E. O., 1986, *Evolution as Entropy. Toward a Unified Theory of Biology*, University of Chicago Press, Chicago.

Brooks, D. R., Thorson, T. B., and Mayes, M. A., 1981, Freshwater stingrays (Potamotrygonidae) and their helminth parasites: testing hypotheses of evolution and coevolution, in: *Advances in Cladistics: Proceedings of the First Meeting of the Willi Hennig Society* (V. A. Funk and D. R. Brooks, eds.), New York Botanical Garden, Bronx, NY, pp. 147–175.

Brooks, J. L., 1984, *Just Before the Origin: Alfred Russell Wallace's Theory of Evolution*, Columbia University Press, New York.

Brown, J. H., and Lomolino, M. V., 1998, *Biogeography*, 2nd Ed., Sinauer, Sunderland, MA.

Brown, J. H., and Maurer, B. A., 1989, Macroecology: the division of food and space among species on continents, *Science* **243**:1143–1150.

Brown, R., 1814, *General Remarks, Geographical and Systematical, on the Botany of Terra Australis*, London.

Browne, J., 1983, *The Secular Ark: Studies in the History of Biogeography*, Yale University Press, New Haven.

Brundin, L. Z., 1988, Phylogenetic biogeography, in: *Analytical Biogeography* (A. A. Myers and P. S. Giller, eds.), Chapman and Hall, New York, pp. 343–369.

Buch, L. von., 1825, *Physicalische Beschreibung der Cancrischen Inseln*, Koeniglichen Akademie der Wissenschaften, Berlin.

Buffon, G. L., 1749–1804, *Histoire Naturelle, Générale et Particulière*, Imprimerie Royale, Puis Plassan, Paris.

Burns, T. P., Patten, B. C., and Higashi, M., 1991, Hierarchical evolution in ecological networks: environs and selection, in: *Theoretical Studies of Ecosystems: The Network Perspective* (M. Higashi and T. P. Burns, eds.), Cambridge University Press, New York, pp. 211–239.

Bush, G. L., 1969, Sympatric host race formation and speciation in frugivorous flies of the genus *Rhagoletis* (Diptera, Tephrytidae), *Evolution* **23**:237–251.

Buss, L. W., 1987, *The Evolution of Individuality*, Princeton University Press, Princeton.

Buzas, M. A., and Culver, S. J., 1994, Species pool and dynamics of marine paleocommunities, *Science* **264**:1439–1441.

Chambers, R., 1844, *Vestiges of the Natural History of Creation*, Churchill, London.

Charlesworth, B., Lande, R., and Slatkin, M., 1982, A neo-Darwinian commentary on macroevolution, *Evolution* **36**:474–498.

Cheetham, A. H., Jackson, J. B. C., and Hayek, L. C., 1994, Quantitative genetics of bryozoan phenotypic evolution. II. Analysis of selection and random change in fossil species using reconstructed genetic parameters, *Evolution* **48**:360–375.

Cohen, A. N., and Carlton, J. T., 1998, Accelerating invasion rate in a highly invaded estuary, *Science* **279**:555–558.

Compston, W., Sambridge, M. S., Reinfrank, R. F., Moczydlowska, M., Vidal, G., and Claesson, S., 1995, Numerical ages of volcanic rocks and the earliest faunal zone within the late Precambrian of east Poland, *J. Geol. Soc., London* **152**:599–611.

Condie, K. C., and Rosen, O. M., 1994, Laurentia–Siberia connection revisited, *Geology* **22**:168–170.

Coope, G. R., 1979, Late Cenozoic Coleoptera: evolution, biogeography, and ecology, *Ann. Revs. Ecol. Syst.* **10**:247–267.

Coope, G. R., 1990, The invasion of northern Europe during the Pleistocene by Mediterranean species of Coleoptera, in: *Biological Invasions in Europe and the Mediterranean Basin* (F. D. Castri, A. J. Hansen and M. Debussche, eds.), Kluwer Academic Publishers, Dordrecht, pp. 203–215.

Cracraft, J., 1988, Deep-history biogeography: retrieving the historical pattern of evolving continental biotas, *Syst. Zool.* **37**:221–236.

Croizat, L., 1958, *Panbiogeography*, Caracas.

Croizat, L., 1964, *Space, Time, and Form, the Biological Synthesis*, Caracas.

Croizat, L., 1982, Vicariance/vicariism, panbiogeography, 'vicariance biogeography', etc.: a clarification, *Syst. Zool.* **31**:291–304.

Croizat, L., Nelson, G., and Rosen, D. E., 1974, Centers of origin and related concepts, *Syst. Zool.* **23**:265–287.

Dalziel, I. W. D., 1991, Pacific margins of Laurentia and East Antarctica–Australia as a conjugate rift pair: evidence and implications for an Eocambrian supercontinent, *Geology* **19**:598–601.

Dalziel, I. W. D., 1997, Neoproterozoic–Paleozoic geography and tectonics: review, hypothesis, and environmental speculations, *Geol. Soc. Amer. Bull.* **109**:16–42.

Damuth, J., 1985, Selection among "species": a formulation in terms of natural functional units, *Evolution* **39**:1132–1146.

Darlington, P. J., Jr., 1959, Area, climate, and evolution, *Evolution* **13**:488–510.

Darwin, C., 1839, *The Voyage of the Beagle*, Penguin Books, New York.

Darwin, C., 1859, *On the Origin of Species by Means of Natural Selection; or the Preservation of Favored Races in the Struggle for Life* (reprinted 1st Ed.), Harvard University Press, Cambridge.

Darwin, C., 1872, *On the Origin of Species by Means of Natural Selection; or the Preservation of Favored Races in the Struggle for Life* (reprinted 6th Ed.), New American Library, New York.

Darwin, C., 1909, *The Foundations of the Origin of Species: Two Essays Written in 1842 and 1844* (F. Darwin, ed.), Cambridge University Press, Cambridge.

Davis, M. B., 1976, Pleistocene biogeography of temperate deciduous forests, *Geosci. Man* **13**:13.

Davis, M., 1986, Climatic instability, time lags, and community disequilibrium, in: *Community Ecology* (J. Diamond and T. Case, eds.), Harper and Row, New York, pp. 269–284.

Dawkins, R., 1976, *The Selfish Gene*, Oxford University Press, Oxford.

Dawkins, R., 1982, *The Extended Phenotype*, W. H. Freeman, San Francisco.

de Candolle, A. P., 1817, Mémoire sur la géographie des plantes de France, considéré dans ses rapports avec la hauteur absolue, *Méms. Phys. Chimie Soc. d'Arcueil* **3**:262–322.

de Candolle, A. P., 1820, Géographie botanique, *Dict. des scies. natur.* **18**:359–422.

de Candolle, A. P., 1821, Elements of the Philosophy of Plants: Containing the Principles of Scientific Botany with a History of the Science and Practical Illustrations, W. Blackwood, Edinburgh.

deMenocal, P., 1995, Plio-Pleistocene African climate, *Science* **270**:53–59.

de Queiroz, K., and Donoghue, M. J., 1988, Phylogenetic systematics and the species problem, *Cladistics* **4**:317–338.

de Queiroz, K., and Donoghue, M. J., 1990, Phylogenetic systematics or Nelson's version of cladistics, *Cladistics* **6**:61–75.

Desmond, A., 1982, *Archetypes and Ancestors*, University of Chicago Press, Chicago.

Dingus, L., 1984, Effects of stratigraphic completeness on interpretations of extinction rates across the Cretaceous–Tertiary boundary, *Paleobiology* **10**:420–438.

Dingus, L., and Sadler, P. M., 1982, The effects of stratigraphic completeness on estimates of evolutionary rates, *Syst. Zool.* **31**:400–412.

Dobzhansky, T., 1937, *Genetics and the Origin of Species*, 1st Ed., Columbia University Press, New York.

Dobzhansky, T., 1951, *Genetics and the Origin of Species*, 3rd Ed., Columbia University Press, New York.

Donoghue, M. J., 1985, A critique of the biological species concept and recommendations for a phylogenetic alternative, *Bryologist* **88**:172–181.

Eldredge, N., 1971, The allopatric model and phylogeny in Paleozoic invertebrates, *Evolution* **25**:156–167.

Eldredge, N., 1979, Alternative approaches to evolutionary theory, *Bull. Carn. Mus. Nat. Hist.* **13**:7–19.

Eldredge, N., 1982, Phenomenological levels and evolutionary rates, *Syst. Zool.* **31**:338–347.

Eldredge, N., 1985a, *Unfinished Synthesis*, Oxford University Press, New York.

Eldredge, N., 1985b, The Ontology of Species, in: *Species and Speciation, Transvaal Museum Monograph No. 4* (E. S. Vrba, ed.), Transvaal Museum, Pretoria, South Africa, pp. 17–20.

Eldredge, N., 1985c, *Time Frames*, Princeton University Press, Princeton.

Eldredge, N., 1986, Information, Economics, and Evolution, *Ann. Revs. Ecol. Syst.* **17**:351–369.

Eldredge, N., 1989a, *Macroevolutionary Dynamics*, McGraw-Hill, New York.

Eldredge, N., 1989b, Punctuated equilibria, rates of change and large-scale entities in evolutionary systems, *J. Soc. Biol. Struct.* **12**:173–184.

Eldredge, N., 1991, *The Miner's Canary*, Prentice-Hall, New York.

Eldredge, N., 1992, *Systematics, Ecology, and the Biodiversity Crisis*, Columbia University Press, New York.

Eldredge, N., 1993, What, if anything, is a species?, in: *Species, Species Concepts, and Primate Evolution* (W. H. Kimbel and L. B. Martin, eds.), Plenum Press, New York, pp. 3–20.

Eldredge, N., 1995, *Reinventing Darwin*, John Wiley & Sons, New York.

Eldredge, N., 1998, *Life in the Balance*, Princeton University Press, Princeton.

Eldredge, N., and Cracraft, J., 1980, *Phylogenetic Patterns and the Evolutionary Process: Method and Theory in Comparative Biology*, Columbia University Press, New York.

Eldredge, N., and Gould, S. J., 1972, Punctuated equilibria: an alternative to phyletic gradualism, in: *Models in Paleobiology* (T. J. Schopf, ed.), Freeman, Cooper, San Francisco, pp. 82–115.

Eldredge, N., and Salthe, S. N., 1984, Hierarchy and evolution, *Oxford Surv. Evol. Biol.* **1**:184–208.

Elton, C. S., 1958, *The Ecology of Invasions by Animals and Plants*, Methuen, London.

Emerson, B. C., Wallis, G. P., and Patrick, B. H., 1997, Biogeographic area relationships in southern New Zealand: a cladistic analysis of Lepidoptera distributions, *J. Biogeog.* **24**:89–99.

Endler, J., 1982, Problems of distinguishing historical from ecological factors in biogeography, *Amer. Zool.* **22**:441–452.

Engelmann, G. F., and Wiley, E. O., 1977, The place of ancestor–descendant relationships in phylogeny reconstruction, *Syst. Zool.* **26**:1–11.

Enghoff, H., 1995, Historical biogeography of the Holarctic: area relationships, ancestral areas, and dispersal of non-marine animals, *Cladistics* **11**:223–263.

Enghoff, H., 1996, Widespread taxa, sympatry, dispersal, and an algorithm for resolved area cladograms, *Cladistics* **12**:349–364.

Ettensohn, F. R., 1985, The Catskill Delta complex and the Acadian Orogeny: a model, in: *The Catskill Delta. Geological Society of America Special Paper 201* (D. L. Woodrow and W. D. Sevon, eds.), Geological Society of America, Boulder, CO, pp. 39–50.

Faith, D. P., 1991, Cladistic permutation tests for monophyly and nonmonophyly, *Syst. Zool.* **40**:366–375.

Faith, D. P., and Trueman, J. W. H., 1996, When the Topology-Dependent Permutation Test (T-PTP) for monophyly returns significant support for monophyly, should that be equated with (a) rejecting a null hypothesis of nonmonophyly, (b) rejecting a null hypothesis of "no structure," (c) failing to falsify a hypothesis of monophyly, or (d) none of the above?, *Syst. Biol.* **45**:580–586.

Farris, J. S., 1988, Hennig86, Port Jefferson Station, NY.

Feynman, R., 1965, *The Character of Physical Law*, MIT Press, Cambridge, MA.

Fichman, M., 1977, Wallace, zoogeography and the problem of land bridges, *J. Hist. Biol.* **10**:45–63.

Fitch, W. M., 1971, Toward defining the course of evolution: minimum change for a specific tree topology, *Syst. Zool.* **20**:406–416.

Flather, C. H., Joyce, L. A., and Bloomgarden, C. A., 1994, Species endangerment patterns in the United States, USDA Forest Service, General Technical Report RM-241.

Flessa, K. W., Cutler, A. H., and Meldahl, K. H., 1993, Time and taphonomy: quantitative estimates of time-averaging and stratigraphic disorder in a shallow marine habitat, *Paleobiology* **19**:266–286.

Forbes, E., 1846, On the connection between the distribution of the existing fauna and flora of the British Isles and the geological changes which have affected their area, especially during the epoch of the Northern Drift, *Mem. Geol. Surv., Gr. Brit.* **1**.

Fortey, R. A., 1984, Global earlier Ordovician transgressions and regressions and their biological implications, in: *Aspects of the Ordovician System* (D. L. Bruton, ed.), Universitetsforlaget Press, Oslo, pp. 37–50.

Fortey, R. A., and Cocks, L. R. M., 1992, The early Paleozoic of the North Atlantic region as a test case for the use of fossils in continental reconstruction, *Tectonophysics* **206**:147–158.

Foster, D. R., Schoonmaker, P. K., and Pickett, S. T. A., 1990, Insights from paleoecology to community ecology, *Tr. Ecol. Evol.* **5**:119–122.

Funk, V. A., and Brooks, D. R., 1990, Phylogenetic systematics as the basis of comparative biology, *Smiths. Contri. Bot.* **73**:1–45.

Gadow, H., 1913, *The Wanderings of Animals*, Cambridge University Press, Cambridge.

Gaffney, E. S., 1979, An introduction to the logic of phylogeny reconstruction, in: *Phylogenetic Analysis and Paleontology* (J. Cracraft and N. Eldredge, eds.), Columbia University Press, New York, pp. 79–111.

Geary, D. H., 1990, Patterns of evolutionary tempo and mode in the radiation of *Melanopsis* (Gastropoda: Melanopsidae), *Paleobiology* **16**:492–511.

Ghiselin, M. T., 1974, A radical solution to the species problem, *Syst. Zool.* **25**:536–544.

Goodwin, B., 1994, *How the Leopard Changed Its Spots*, Charles Scribner's & Sons, New York.

Gould, S. J., 1965, Is uniformitarianism necessary?, *Am. J. Sci.* **263**:223–228.

Gould, S. J., 1978, The Great Scablands debate, *Nat. Hist.* **87**:12–18.

Gould, S. J., 1980, Is a new and general theory of evolution emerging?, *Paleobiology* **6**:119–130.

Gould, S. J., 1982, Darwinism and the expansion of evolutionary theory, *Science* **216**:380–387.

Gould, S. J., 1987, *Time's Arrow, Time's Cycle*, Harvard University Press, Cambridge.

Gould, S. J., 1989, *Wonderful Life*, W. W. Norton, New York.

Gould, S. J., 1990, Speciation and sorting as the source of evolutionary trends, or "things are seldom what they seem," in: *Evolutionary Trends* (K. J. McNamara, ed.), Belhaven Press, London, pp. 3–27.

Gould, S. J., 1996, Mr. Sophia's Pony, *Nat. Hist.* **105**:20–24, 66–69.

Gould, S. J., and Eldredge, N., 1977, Punctuated equilibria: the tempo and mode of evolution reconsidered, *Paleobiology* **3**:115–151.

Graham, R. W., 1986, Response of mammalian communities to environmental changes during the late Quaternary, in: *Community Ecology* (J. Diamond and T. J. Case, eds.), Harper and Row, New York, pp. 300–313.

Graham, R. W., 1992, Late Pleistocene faunal changes as a guide to understanding effects of greenhouse warming on the mammalian fauna of North America, in: *Global Warming and Biological Diversity* (R. L. Peters and T. E. Lovejoy, eds.), Yale University Press, New Haven, pp. 76–87.

Graham, R. W., Jr., Graham, M. A., Schroeder, E. K., III, Anderson, E., Barnosky, A. D., Burns, J. A., Churcher, C. S., Grayson, D. K., Guthrie, R. D., Harington, C. R., Jefferson, G. T., Martin, L. D., McDonald, H. G., Morlan, R. E., Jr., Webb, S. D., Werdelin, L., and Wilson, M. C., 1996, Spatial response of mammals to late Quaternary environmental fluctuations, *Science* **272**:1601–1606.

Grinnell, G., 1974, The rise and fall of Darwin's first theory of transmutation, *J. Hist. Biol.* **7**:259–273.

Gulick, J. T., 1888, Divergent evolution through cumulative segregation, *J. Linn. Soc., Zool.* **20**:189–274.

Hallam, A., 1967, The bearing of certain paleogeographic data on continental drift, *Palaeogeog. Palaeoclim. Palaeoecol.* **3**:201–224.

Hallam, A., 1977, Jurassic bivalve biogeography, *Paleobiology* **3**:58–73.

Hallam, A., 1981, *Great Geological Controversies*, Oxford University Press, New York.

Hallam, A., 1983, Early and mid-Jurassic molluscan biogeography and the establishment of the central Atlantic seaway, *Palaeogeog. Palaeoclim. Palaeoecol.* **43**:181–193.

Hallam, A., 1992, *Phanerozoic Sea-Level Changes*, Columbia University Press, New York.

Hallam, A., 1994, *An Outline of Phanerozoic Biogeography*, Vol. 10, Oxford University Press, Oxford.

Harold, A. S., and Mooi, R. D., 1994, Areas of endemism: definition and recognition criteria, *Syst. Biol.* **43**:261–266.

Harvey, P. H., and Pagel, M. D., 1991, *The Comparative Method in Evolutionary Biology*, Oxford University Press, Oxford.

Hays, J. D., Imbrie, J., and Shackleton, N. J., 1976, Variations in the Earth's orbit: pacemaker of the ice ages, *Science* **194**:1121–1132.

Henderson, R. A., and Heron, M. L., 1977, A probabilistic method of paleobiogeographic analysis, *Lethaia* **10**:1–16.

Hennig, W., 1966, *Phylogenetic Systematics*, University of Illinois Press, Urbana.

Hillis, D. M., 1991, Discriminating between phylogenetic signal and random noise in DNA sequences, in: *Phylogenetic Analysis of DNA Sequences* (M. M. Miyamoto and J. Cracraft, eds.), Oxford University Press, New York, pp. 278–294.

Hoffman, P. H., 1991, Did the breakout of Laurentia turn Gondwanaland inside out?, *Science* **252**:1409–1413.

Holland, S. M., 1996, Recognizing artifactually generated coordinated stasis: implications of numerical models and strategies for field tests, *Palaeogeog. Palaeoclim. Palaeoecol.* **127**:147–156.

Holterhoff, P., 1996, Crinoid biofacies in Upper Carboniferous cyclothems, midcontinent North America: faunal tracking and the role of regional processes in biofacies recurrence, *Palaeogeog. Palaeoclim. Palaeoecol.* **127**:47–82.

Hooker, J. D., 1853, *The Botany of the Antarctic Voyage of H. M. Discovery Ships "Erebus" and "Terror" in the Years 1839–1843. II. Flora Novae-Zelandiae. Part I. Flowering Plants*, Lovell Reeve, London.

Hovenkamp, P., 1997, Vicariance events, not areas, should be used in biogeographical analysis, *Cladistics* **13**:67–79.

Hull, D. L., 1976, Are species really individuals?, *Syst. Zool.* **25**:174–191.

Hull, D. L., 1978, A matter of individuality, *Phil. Sci.* **45**:335–360.

Hull, D. L., 1980, Individuality and selection, *Ann. Revs. Ecol. Syst.* **11**:311–332.

Hull, D. L., 1988, *Science as a Process*, University of Chicago Press, Chicago.

Humboldt, A. von, 1816, On the laws observed in the distribution of vegetable forms, *Phil. Mag. J.* **47**:446.

Humboldt, A. von, 1820, Géographie des plantes, *Dict. Scies. Nat.* **18**:422–432.

Humphries, C. J., and Parenti, L., 1986, Cladistic biogeography, *Oxford Mon. Biogeog.* **2**:1–98.

Humphries, C. J., Ladiges, P. Y., Roos, M., and Zandee, M., 1988, Cladistic Biogeography, in: *Analytical Biogeography* (A. A. Myers and P. S. Giller, eds.), Chapman and Hall, New York City, pp. 371–404.

Huntley, B., and Webb, T., III, 1989, Migration: species' response to climatic variations caused by changes in the earth's orbit, *J. Biogeog.* **16**:5–19.

Huxley, T. H., 1870, Anniversary address, in: *The Scientific Memoirs of Thomas Henry Huxley* (M. Foster and E. R. Lankester, eds.), Macmillan, London, pp. 510–550.

Imbrie, J., and Imbrie, J. Z., 1980, Modeling the climatic response to orbital variations, *Science* **207**:943–953.

Jablonski, D., 1991, Extinctions: a paleontological perspective, *Science* **253**:754–757.

Jablonski, D., Flessa, K. W., and Valentine, J. W., 1985, Biogeography and paleontology, *Paleobiology* **11**:75–90.

Jackson, J. B. C., 1992, Pleistocene perspectives on coral reef community structure, *Am. Zool.* **32**:719–730.

Jackson, J. B. C., and Cheetham, A. H., 1990, Evolutionary significance of morphospecies: a test with cheilostome Bryozoa, *Science* **248**:579–583.

Jackson, J. B. C., Budd, A. F., and Pandolfi, J. M., 1996, The shifting balance of natural communities?, in: *Evolutionary Paleobiology* (D. Jablonski, D. H. Erwin and J. H. Lipps, eds.), University of Chicago Press, Chicago, pp. 89–122.

Jell, P. A., 1974, Faunal provinces and possible planetary reconstruction of the Middle Cambrian, *J. Geol.* **82**:319–350.

Jordan, D. S., 1908, The law of geminate species, *Am. Nat.* **42**:73–80.

Kent, D. V., 1985, Paleocontinental setting for the Catskill Delta, in: *The Catskill Delta. Geological Society of America Special Paper 201* (D. L. Woodrow and W. D. Sevon, eds.), Geological Society of America, Boulder, CO, pp. 9–14.

Kinch, M. P., 1980, Geographical distribution and the origin of life: the development of early nineteenth century British explanations, *J. Hist. Biol.* **13**:91–119.

Kirchgasser, W. T., Oliver, W. A., Jr., and Rickard, L. V., 1985, Devonian series boundaries in the eastern United States, *Cour. Forsch. Senckenberg.* **75**:233–260.

Kluge, A. G., 1988, Parsimony in vicariance biogeography: a quantitative method and a Greater Antillean example, *Syst. Zool.* **37**:315–328.

Knoll, A. H., 1991, End of the Proterozoic eon, *Sci. Amer.* **265**:64–73.

Knoll, A. H., 1996, Daughter of time, *Paleobiology* **22**:1–7.

Knoll, A. H., Bambach, R. K., Canfield, D. E., and Grotzinger, J. P., 1996, Comparative earth history and late Permian mass extinction, *Science* **273**:452–457.

Kottler, M. J., 1978, Charles Darwin's biological species concept and theory of geographic speciation: the transmutation notebooks, *Ann. Sci.* **35**:275–297.

Lamarck, J.-B., 1809, *Philosophie Zoologique*, Paris.

Latreille, P. A., 1822, Introduction a la géographie générale des Arachnides et des Insectes, *Edinburgh N. Phil. J.* **5/6**:370–378, 51–62.

Lich, D., 1990, *Cosomys primus*: a case for stasis, *Paleobiology* **16**:384–395.

Lieberman, B. S., 1992, An extension of the SMRS concept into a phylogenetic context, *Evol. Th.* **10**:157–161.

Lieberman, B. S., 1994, Evolution of the trilobite subfamily Proetinae Salter, 1864 and the origin, diversification, evolutionary affinity and extinction of the Middle Devonian proetid fauna of Eastern North America, *Bull. Amer. Mus. Nat. Hist.* **223**:1–176.

Lieberman, B. S., 1995, Phylogenetic trends and speciation: analyzing macroevolutionary processes and levels of selection, in: *New Approaches to Speciation in the Fossil Record* (D. H. Erwin and R. L. Anstey, eds.), Columbia University Press, New York, pp. 316–337.

Lieberman, B. S., 1997, Early Cambrian paleogeography and tectonic history: a biogeographic approach, *Geology* **25**:1039–1042.

Lieberman, B. S., 1998, Cladistic analysis of the Early Cambrian olenelloid trilobites, *J. Paleontol.* **72**:59–78.

Lieberman, B. S., 1999a, Testing the Darwinian legacy of the Cambrian radiation using trilobite phylogeny and biogeography, *J. Paleontol.* **73**:176–181.

Lieberman, B. S., 1999b, Systematic revision of the Olenelloidea (Trilobita, Cambrian), *Bull. Yale U. Peabody Mus. Nat. Hist.* **45**:1–150.

Lieberman, B.S., 2000. A probabilistic analysis of rates of evolution during the Cambrian radiation. *PNAS, U.S.A.*, in press.

Lieberman, B. S., and Dudgeon, S., 1996, An evaluation of stabilizing selection as a mechanism for stasis, *Palaeogeog. Palaeoclim. Palaeoecol.* **127**:229–238.

Lieberman, B. S., and Eldredge, N., 1996, Trilobite biogeography in the Middle Devonian: geological processes and analytical methods, *Paleobiology* **22**:66–79.

Lieberman, B. S., and Kloc, G. J., 1997, Evolutionary and biogeographic patterns in the Asteropyginae (Trilobita, Devonian) Delo, 1935, *Bull. Amer. Mus. Nat. Hist.* **232**:1–127.

Lieberman, B. S., and Vrba, E. S., 1995, Hierarchy theory, selection, and sorting: a phylogenetic perspective, *BioSci.* **45**:394–399.

Lieberman, B. S., Allmon, W. D., and Eldredge, N., 1993, Levels of selection and macroevolutionary patterns in the turritellid gastropods, *Paleobiology* **19**:205–215.

Lieberman, B. S., Brett, C. E., and Eldredge, N., 1994, Patterns and processes of stasis in two species lineages of brachiopods from the Middle Devonian of New York State, *Am. Mus. Nat. Hist. Nov.* **3114**:1–23.

Lieberman, B. S., Brett, C. E., and Eldredge, N., 1995, A study of stasis and change in two species lineages from the Middle Devonian of New York State, *Paleobiology* **21**:15–27.

Liebherr, J. K., 1988, General patterns in West Indian insects, and graphical biogeographic analysis of some circum-Caribbean *Platynus* beetles (Carabidae), *Syst. Zool.* **37**:385–409.

Lloyd, E. A., and Gould, S. J., 1993, Species selection on variability, *Proc. Nat. Acad. Sci., USA* **90**:595–599.

Lydeard, C., Wooten, M. C., and Meyer, A., 1995, Molecules, morphology, and area cladograms: a cladistic and biogeographic analysis of *Gambusia* (Teleostei: Poeciliidae), *Syst. Biol.* **44**:221–236.

Lydekker, R., 1896, *A Geographical History of Mammals*, Cambridge University Press, Cambridge.

Lyell, C., 1831, *Principles of Geology* (1st Ed.), Vol. 1, University of Chicago Press, Chicago.

Lyell, C., 1832, *Principles of Geology* (1st Ed.), Vol. 2, University of Chicago Press, Chicago.

Lynch, J. D., 1989, The gauge of speciation: on the frequencies of modes of speciation, in: *Speciation and Its Consequences* (D. Otte and J. A. Endler, eds.), Sinauer, Sunderland, MA, pp. 527–553.

MacArthur, R. H., and Wilson, E. O., 1967, *The Theory of Island Biogeography*, Princeton University Press, Princeton.

Maddison, W. P., and Maddison, D. R., 1992, *MacClade: Analysis of phylogeny and character evolution. Version 3.04*, Sinauer Associates, Sunderland, MA.

Marshall, C. R., 1990, Confidence intervals on stratigraphic ranges, *Paleobiology* **16**:1–10.

Matthew, W. D., 1915, Climate and Evolution, *Ann. N. Y. Acad. Sci.* **24**:171–318.

Matthew, W. D., 1939, *Climate and Evolution* (2nd Ed.), New York Academy of Sciences, New York.

Mayden, R. L., 1988, Vicariance biogeography, parsimony, and evolution in North American freshwater fishes, *Syst. Zool.* **37**:329–355.

Mayr, E., 1942, *Systematics and the Origin of Species*, Dover Press, New York.

Mayr, E., 1963, *Animal Species and Evolution*, Harvard University Press, Cambridge.

Mayr, E., 1976, *Evolution and the Diversity of Life: Selected Essays*, Harvard University Press, Cambridge.

Mayr, E., 1982, *The Growth of Biological Thought*, Harvard University Press, Cambridge.

McCune, A. R., 1987, Lakes as laboratories of evolution: endemic fishes and environmental cyclicity, *Palaios* **2**:446–454.

McGhee, G. R., Jr., 1988, The Late Devonian extinction event: evidence for abrupt ecosystem collapse, *Paleobiology* **14**:250–257.

McGhee, G. R., Jr., 1989, The Frasnian-Famennian extinction event, in: *Mass Extinctions: Processes and Evidence* (S. K. Donovan, ed.), Columbia University Press, New York, pp. 133–151.

McGhee, G. R., Jr., 1996, *The Late Devonian mass extinction*, Columbia University Press, New York.

McGowan, J. A., Cayan, D. R., and Dorman, L. M., 1998, Climate–ocean variability and ecosystem response in the northeast Pacific, *Science* **281**:210–217.

McKenna, M. C., 1975, Fossil mammals and Early Eocene North Atlantic land continuity, *Ann. Miss. Bot. Gard.* **62**:335–353.

McKenna, M. C., 1983, Holarctic landmass rearrangement, cosmic events, and Cenozoic terrestrial organisms, *Ann. Miss. Bot. Gard.* **70**:459–489.

McKerrow, W. S., Scotese, C. R., and Brasier, M. D., 1992, Early Cambrian continental reconstructions, *J. Geol. Soc., London* **149**:599–606.

McMenamin, M. A., and McMenamin, D. L., 1990, *The Emergence of Animals*, Columbia University Press, New York.

Meldahl, K. H., Flessa, K. W., and Cutler, A. H., 1997, Time-averaging and postmortem skeletal survival in benthic fossil assemblages: quantitative comparisons among Holocene environments, *Paleobiology* **23**:207–229.

Michaux, B., 1991, Distributional patterns and tectonic development in Indonesia: Wallace reinterpreted, *Austral. Syst. Bot.* **4**:25–36.

Michaux, B., 1996, The origin of southwest Sulawesi and other Indonesian terranes: a biological view, *Palaeogeog. Palaeoclim. Palaeoecol.* **122**:167–183.

Mickevich, M. F., 1981, Quantitative phylogenetic biogeography, in: *Advances in Cladistics: Proceedings of the First Meeting of the Willi Hennig Society* (V. A. Funk and D. R. Brooks, eds.), New York Botanical Garden, Bronx, NY, pp. 209–222.

Miller, A. I., and Mao, S., 1995, Association of orogenic activity with the Ordovician radiation of marine life, *Geology* **23**:305–308.

Miyamoto, M. M., 1985, Consensus cladograms and general classifications, *Cladistics* **1**:186–189.

Morris, P. J., Ivany, L. C., Schopf, K. M., and Brett, C. E., 1995, The challenge of paleoecological stasis: reassessing sources of evolutionary stability, *Proc. Nat. Acad. Sci., USA* **92**:11269–11273.

Morrone, J. J., 1994, On the identification of areas of endemism, *Syst. Biol.* **43**:438–441.

Morrone, J. J., and Carpenter, J. M., 1994, In search of a method for cladistic biogeography: an empirical comparison of components analysis, Brooks Parsimony Analysis, and three-area statements, *Cladistics* **10**:99–153.

Morrone, J. J., and Crisci, J. V., 1995, Historical biogeography: introduction to methods, *Ann. Revs. Ecol. Syst.* **26**:373–401.

Murray, A., 1866, *The Geographical Distribution of Mammals*, Day, London.

Nelson, G., 1976, Biogeography, the vicariance paradigm, and continental drift, *Syst. Zool.* **24**:490–504.

Nelson, G., 1978, From Candolle to Croizat: comments on the history of biogeography, *J. Hist. Biol.* **11**:269–305.

Nelson, G., 1989, Cladistics and evolutionary models, *Cladistics* **5**:275–289.

Nelson, G., and Platnick, N. I., 1981, *Systematics and Biogeography: Cladistics and Vicariance*, Columbia University Press, New York.

Nixon, K. C., and Wheeler, Q. D., 1992, Extinction and the origin of species, in: *Extinction and Phylogeny* (M. J. Novacek and Q. D. Wheeler, eds.), Columbia University Press, New York, pp. 119–143.

Noonan, G. R., 1988, Biogeography of North American and Mexican insects, and a critique of vicariance biogeography, *Syst. Zool.* **37**:366–384.

Norell, M. A., and Novacek, M. J., 1992, The fossil record and evolution: comparing cladistic and paleontologic evidence for vertebrate history, *Science* **255**:1690–1693.

Oliver, W. A., Jr., 1977, Biogeography of late Silurian and Devonian rugose corals, *J. Paleontol.* **50**:365–373.

Oliver, W. A., Jr., 1990, Extinctions and migrations of Devonian rugose corals in the Eastern Americas Realm, *Lethaia* **23**:167–178.

Oliver, W. A., Jr., and Pedder, A. E. H., 1989, Origins, migrations, and extinctions of Devonian Rugosa on the North American plate, *Mems. Assoc. Austral. Palaeontol.* **8**:231–237.

Olsen, P. E., 1984, Periodicity of lake-level cycles in the Late Triassic Lockatong Formation of the Newark Basin (Newark Supergroup, New Jersey and Pennsylvania), in: *Milankovitch and Climate: NATO Symposium, Part 1* (A. Berger, J. Imbrie, J. Hays, G. Kukla and B. Saltzman, eds.), Reidel, Dordrecht, pp. 129–146.

Olsen, P. E., 1986, A 40-million-year lake record of early Mesozoic orbital climatic forcing, *Science* **234**:842–848.

Oosterzee, P., Van., 1997, *Where Worlds Collide*, Cornell University Press, Ithaca.

Page, R. D. M., 1990, Component analysis: a valiant failure?, *Cladistics* **6**:119–136.

Page, R. D. M., and Lydeard, C., 1994, Towards a cladistic biogeography of the Caribbean, *Cladistics* **10**:21–41.

Pandolfi, J. M., 1996, Limited membership in Pleistocene reef coral assemblages from the Huon Peninsula, Papua New Guinea: constancy during global change, *Paleobiology* **22**:152–176.

Paterson, H. E., 1985, The recognition concept of species, in: *Species and Speciation* (E. S. Vrba, ed.), Transvaal Museum Monographs, pp. 21–29.

Patterson, C., 1983, Aims and methods in biogeography, in: *Evolution, Time and Space: The Emergence of the Biosphere* (R. W. Sims, J. H. Price and P. E. S. Whalley, eds.), Academic Press, New York, pp. 1–28.

Patterson, C., and Smith, A. B., 1987, Is the periodicity of extinctions a taxonomic artefact?, *Nature* **330**:248–252.

Patzkowsky, M. E., and Holland, S. M., 1997, Patterns of turnover in Middle and Upper Ordovician brachiopods of the eastern United States: a test of coordinated stasis, *Paleobiology* **23**:420–443.

Pelechaty, S., 1996, Stratigraphic evidence for the Siberia–Laurentia connection and Early Cambrian rifting, *Geology* **24**:719–722.

Platnick, N. I., 1976, Concepts of dispersal in historical biogeography, *Syst. Zool.* **25**:294–295.

Platnick, N. I., and Gaffney, E. S., 1977, Systematics: a Popperian perspective (Reviews of "The Logic of Scientific Discovery" and "Conjectures and Refutations" by Karl R. Popper), *Syst. Zool.* **26**:360–365.

Platnick, N. I., and Gaffney, E. S., 1978*a*, Evolutionary biology: a Popperian perspective, *Syst. Zool.* **27**:137–141.

Platnick, N. I., and Gaffney, E. S., 1978*b*, Systematics and the Popperian paradigm, *Syst. Zool.* **27**:381–388.

Platnick, N. I., and Nelson, G., 1978, A method of analysis for historical biogeography, *Syst. Zool.* **27**:1–16.

Prichard, J. C., 1826, *Researches into the Physical History of Mankind* (2nd Ed.), J. & A. Arch, London.

Prothero, D. R., 1994, *The Eocene–Oligocene Transition*, Columbia University Press, New York.

Raup, D. M., 1976, Species diversity in the Phanerozoic: an interpretation, *Paleobiology* **2**:289–297.

Raup, D. M., 1986, *The Nemesis Affair*, W. W. Norton, New York.

Raup, D. M., and Sepkoski, J. J., Jr., 1986, Periodic extinction of families and genera, *Science* **231**:833–836.

Raxworthy, C. J., and Nussbaum, R. A., 1996, Patterns of endemism for terrestrial vertebrates in eastern Madagascar, *Biogéog. Madagas.* **1996**:369–383.

Ree, R. H., and Donoghue, M. J., 1998, Step matrices and the interpretation of homoplasy, *Syst. Biol.* **47**:582–588.

Renne, P. R., Zichao, Z., Richards, M. A., Black, M. T., and Basu, A. R., 1995, Synchrony and causal relations between Permian–Triassic boundary crises and Siberian flood volcanism, *Science* **269**:1413–1416.

Richardson, R. A., 1981, Biogeography and the genesis of Darwin's ideas on transmutation, *J. Hist. Biol.* **14**:1–41.

Ricklefs, R. E., 1979, *Ecology* (2nd Ed.), Chiron Press, New York.

Ricklefs, R. E., 1987, Community diversity: relative roles of local and regional processes, *Science* **253**:167–171.

Ronquist, F., 1994, Ancestral areas and parsimony, *Syst. Biol.* **43**:267–274.

Ronquist, F., 1995, Ancestral areas revisited, *Syst. Biol.* **44**:572–575.

Ronquist, F., 1997, Dispersal-vicariance analysis: a new approach to the quantification of historical biogeography, *Syst. Biol.* **46**:195–203.

Ronquist, F., 1998, Phylogenetic approaches in coevolution and biogeography, *Zool. Scr.* **26**:313–322.

Rosen, B. R., 1988, From fossils to earth history: applied historical biogeography, in: *Analytical Biogeography* (A. A. Myers and P. S. Giller, eds.), Chapman and Hall, London, pp. 437–478.

Rosen, B., 1991, *Life Itself*, Columbia University Press, New York.

Rosen, D. E., 1978, Vicariant patterns and historical explanation in biogeography, *Syst. Zool.* **27**:159–188.

Rosen, D. E., 1979, Fishes from the uplands and intermontane basins of Guatemala: revisionary studies and comparative geography, *Bull. Amer. Mus. Nat. Hist.* **162**:269–375.

Ross, H. H., 1972, The origin of species diversity in ecological communities, *Taxon* **21**:253–259.

Ross, H. H., 1986, Resource partitioning in fish assemblages: a review of field studies, *Copeia* **86**:352–388.

Rowell, A. J., McBride, D. J., and Palmer, A. R., 1973, Quantitative study of Trempealeauian (latest Cambrian) trilobite distribution in North America, *Bull. Geol. Soc. Amer.* **84**:3429–3442.

Sadler, P. M., 1981, Sediment accumulation rates and the completeness of stratigraphic sections, *J. Geol.* **89**:569–584.

Salthe, S. N., 1985, *Evolving Hierarchical Systems*, Columbia University Press, New York.

Schindel, D. E., 1980, Microstratigraphic sampling and the limits of paleontologic resolution, *Paleobiology* **6**:408–426.

Schluter, D., Price, T., Mooers, A. Ø., and Ludwig, D., 1997, Likelihood of ancestor states in adaptive radiation, *Evolution* **51**:1699–1711.

Schultz, T. R., Cocroft, R. B., and Churchill, G. A., 1996, The reconstruction of ancestral character states, *Evolution* **50**:504–511.

Sclater, P. L., 1857, On the general geographical distribution of the members of Class Aves, *J. Linn. Soc. London, Zool.* **2**:130–145.

Scotese, C. R., 1997, *Paleogeographic Atlas*, PALEOMAP Project, University of Texas at Arlington, Arlington, TX.

Scotese, C. R., and McKerrow, W. S., 1990, Revised world maps and introduction, in: *Palaeozoic Palaeogeography and Biogeography, Geological Society (London) Memoir 2* (W. S. McKerrow and C. R. Scotese, eds.), Geological Society of London, London, pp. 1–21.

Sheldon, P. R., 1987, Parallel gradualistic evolution of Ordovician trilobites, *Nature* **330**:561–563.

Siddall, M. E., 1996, Phylogenetic covariance probability: confidence and historical associations, *Syst. Biol.* **45**:48–66.

Signor, P. W., and Lipps, J. H., 1992, Origin and early radiation of the Metazoa, in: *Origin and Early Evolution of the Metazoa* (J. H. Lipps and P. W. Signor, eds.), Plenum Press, New York, pp. 3–23.

Simberloff, D., 1987, Calculating the probabilities that cladograms match: a method of biogeographic inference, *Syst. Zool.* **36**:175–195.

Simpson, G. G., 1944, *Tempo and Mode in Evolution*, Columbia University Press, New York.

Simpson, G. G., 1961, *Principles of Animal Taxonomy*, Columbia University Press, New York.

Simpson, G. G., 1965, *The Geography of Evolution*, Chilton Press, New York, NY.

Smith, A. B., 1994, *Systematics and the Fossil Record: Documenting Evolutionary Patterns*, Blackwell Scientific, Cambridge, MA.

Smith, A. B., and Jeffery, C. H., 1998, Selectivity of extinction among sea urchins at the end of the Cretaceous period, *Nature* **392**:69–71.

Smuts, J. C., 1925, *Holism and Evolution*, Viking Press, New York.

Sober, E., 1984, *The Nature of Selection*, MIT Press, Cambridge.

Sober, E., 1988, *Reconstructing the Past*, MIT Press, Cambridge.

Soest, R. W. M. von, and Hajdu, E., 1997, Marine area relationships from twenty sponge phylogenies. A comparison of methods and coding strategies, *Cladistics* **13**:1–20.

Sokal, R. R., and Crovello, T. J., 1970, The biological species concept: a critical evaluation, *Amer. Nat.* **104**:127–153.

Soper, N. J., Strachan, R. A., Holdsworth, R. E., Gayer, R. A., and Greiling, R. O., 1992, Sinistral transpression and the Silurian closure of Iapetus, *J. Geol. Soc., London* **14**:871–880.

Stanley, S. M., 1987, *Extinction*, Scientific American Books, New York.

Stanley, S. M., 1998, *Earth System History*, W. H. Freeman and Co., San Francisco.

Stanley, S. M., and Yang, X., 1987, Approximate evolutionary stasis for bivalve morphology over millions of years: a multivariate, multilineage study, *Paleobiology* **13**:113–139.

Stanley, S. M., and Yang, X., 1994, A double mass extinction at the end of the Paleozoic era, *Science* **266**:1340–1344.

Stevens, G., 1992, Spilling over the competitive limits to species coexistence, in: *Systematics, Ecology, and the Biodiversity Crisis* (N. Eldredge, ed.), Columbia University Press, New York, pp. 40–58.

Stolzenburg, W., 1999, The evil has landed: invasive species targeted at federal level, *Nature Conserv.* **49**:8.

Streidter, G. F., and Northcutt, R. G., 1991, Biological hierarchies and the concept of homology, *Br., Behav., and Evol.* **38**:177–189.

Sulloway, F. J., 1979, Geographic isolation in Darwin's thinking: the vicissitudes of a crucial idea, in: *Studies in the History of Biology* (W. Coleman and C. Limoges, eds.), Johns Hopkins University Press, Baltimore, pp. 23–65.

Swett, K., 1981, Cambro-Ordovician strata in Ny Friesland, Spitsbergen and their paleotectonic significance, *Geol. Mag.* **118**:225–250.

Swofford, D. L., 1993, *PAUP (Phylogenetic analysis using parsimony) version 3.1.1*, Sinauer Associates, Sunderland, MA.

Swofford, D. L., 1998, *PAUP (Phylogenetic analysis using parsimony), version 4.0*, Sinauer Associates, Sunderland, MA.

Torsvik, T. H., Smethurst, M. A., Meert, J. G., Van der Voo, R., McKerrow, W. S., Brasier, M. D., Sturt, B. A., and Walderhaug, H. J., 1996, Continental break-up and collision in the Neoproterozoic and Paleozoic—a tale of Baltica and Laurentia, *Earth-Sci. Revs.* **40**:229–258.

Uyeda, S., 1978, *The New View of the Earth*, W. H. Freeman and Co., San Francisco.

Valentine, J. W., 1989, How good was the fossil record? Clues from the Californian Pleistocene, *Paleobiology* **15**:83–94.

Valentine, J. W., and May, C. L., 1996, Hierarchies in biology and paleontology, *Paleobiology* **22**:23–33.

Valentine, J. W., and Moores, E. M., 1970, Plate tectonic regulation of faunal diversity and sea level: a model, *Nature* **228**:657–659.

Valentine, J. W., and Moores, E. M., 1972, Global tectonics and the fossil record, *J. Geol.* **80**:167–184.

Valentine, J. W., Foin, T. C., and Peart, D., 1978, A provincial model of Phanerozoic marine diversity, *Paleobiology* **4**:55–66.

Van Valkenburg, B., 1988, Trophic diversity in past and present guilds of large predatory mammals, *Paleobiology* **14**:155–173.

Vermeij, G., 1978, *Biogeography and Adaptation*, Harvard University Press, Cambridge.

Vrba, E. S., 1980, Evolution, species and fossils: how does life evolve?, *S. Afr. J. Sci.* **76**:61–84.

Vrba, E. S., 1983, Macroevolutionary trends: new perspectives on the roles of adaptation and incidental effect, *Science* **221**:387–389.

Vrba, E. S., 1984a, What is species selection? *Syst. Zool.* **33**:318–328.

Vrba, E. S., 1984b, Evolutionary pattern and process in the sister-group Alcelaphini-Aepycerotini (Mammalia: Bovidae), in: *Living Fossils* (N. Eldredge and S. M. Stanley, eds.), Springer-Verlag, New York, pp. 62–79.

Vrba, E. S., 1985, Environment and evolution: alternative causes of the temporal distribution of evolutionary events, *S. Afr. J. Sci.* **81**:229–236.

Vrba, E. S., 1989, Levels of selection and sorting with special reference to the species level, *Oxford Surv. Evol. Biol.* **6**:111–168.

Vrba, E. S., 1992, Mammals as a key to evolutionary theory, *J. Mammal.* **73**:1–28.

Vrba, E. S., 1993, Turnover-pulses, the Red Queen, and related topics, *Amer. J. Sci.* **293**:418–452.

Vrba, E. S., 1995, Species as habitat-specific, complex systems, in: *Speciation and the Recognition Concept* (D. M. Lambert and H. G. Spencer, eds.), Johns Hopkins University Press, Baltimore, pp. 3–44.

Vrba, E. S., 1996, On the connection between paleoclimate and evolution, in: *Paleoclimate and Evolution with Emphasis on Human Origins* (E. S. Vrba, G. H. Denton, T. C. Partridge and L. H. Burckle, eds.), Yale University Press, New Haven, pp. 24–45.

Vrba, E. S., and Eldredge, N., 1984, Individuals, hierarchies and processes: toward a more complete evolutionary theory, *Paleobiology* **10**:146–171.

Vrba, E. S., and Gould, S. J., 1986, The hierarchical expansion of sorting and selection: sorting and selection cannot be equated, *Paleobiology* **12**:217–228.

Waggoner, B. M., 1996, Phylogenetic hypotheses of the relationships of arthropods to Precambrian and Cambrian problematic fossil taxa, *Syst. Biol.* **45**:190–222.

Wagner, M., 1868, *Die Darwinishce Theorie und das Migrationgesetz der Organismen*, Duncker & Humblot, Leipzig.

Wagner, M., 1889, *Die Entstehung der Arten durch räumliche Sonderung: Gesammelte Aufsätze*, Benno Schwabe, Basel.

Wallace, A. R., 1852, On the monkeys of the Amazon, *Proc. Zool. Soc. London* **20**:107–110.

Wallace, A. R., 1855, On the law which has regulated the introduction of new species, *Ann. Mag. Nat. Hist., 2nd Ser.* **16**:184–196.

Wallace, A. R., 1857, On the natural history of the Aru Islands, *Ann. Mag. Nat. Hist., 2nd Ser.* **20**:473–485.

Wallace, A. R., 1860, On the zoological geography of the Malay Archipelago, *J. Proc. Linn. Soc., Zool.* **4**:172–184.

Wallace, A. R., 1863, On the physical geography of the Malay Archipelago, *J. Roy. Geogr. Soc.* **33**:217–234.

Wallace, A. R., 1869, *Malay Archipelago* (Reprint of 10th Ed.), Dover, New York.

Wallace, A. R., 1876, *The Geographical Distribution of Animals*, Harpers, New York.

Wallace, A. R., 1880, *The Malay Archipelago* (7th Ed.), Macmillan, London.

Wallace, A. R., 1889, *Darwinism: An Exposition of the Theory of Natural Selection with Some of Its Applications*, Macmillan, London.

Watkins, T. H., 1996, The greening of the empire: Sir Joseph Banks, *Nat. Geog.* **190**:28–53.

Webb, S. D., 1978, A history of savanna vertebrates in the new world. part II: South America and the Great Interchange, *Ann. Revs. Ecol. Syst.* **9**:393–426.

Webb, S. D., 1989, The fourth dimension in North American terrestrial mammal communities, in: *Patterns in the Structure of Mammalian Communities* (D. W. Morris, ed.), Texas Tech. University Museum, Lubbock, TX, pp. 181–203.

Wells, M. L., Vallis, G. K., and Silver, E. A., 1999, Tectonic processes in Papua New Guinea and past productivity in the eastern equatorial Pacific Ocean, *Nature* **398**:601–604.

Westrop, S. R., 1996, Temporal persistence and stability of Cambrian biofacies: Sunwaptan (Upper Cambrian) trilobite faunas of North America, *Palaeogeog. Palaeoclim. Palaeoecol.* **127**: 33–46.

Whewell, W., 1840, *The Philosophy of the Inductive Sciences*, Parker, London.

Whittington, H. B., and Hughes, C. P., 1972, Ordovician geography and faunal provinces deduced from trilobite distribution, *Phil. Trans. Roy. Soc. London, Ser. B* **263**:235–278.

Wiley, E. O., 1975, Karl R. Popper, systematics, and classification: a reply to Walter Bock and other evolutionary taxonomists, *Syst. Zool.* **24**:233–242.

Wiley, E. O., 1978, The evolutionary species concept revisited, *Syst. Zool.* **27**:17–26.

Wiley, E. O., 1981, *Phylogenetics: The Theory and Practice of Phylogenetic Systematics*, John Wiley & Sons, New York.

Wiley, E. O., 1988*a*, Parsimony analysis and vicariance biogeography, *Syst. Zool.* **37**:271–290.

Wiley, E. O., 1988*b*, Vicariance biogeography, *Ann. Revs. Ecol. Syst.* **19**:513–542.

Wiley, E. O., and Mayden, R. L., 1985, Species and speciation in phylogenetic systematics, with examples from the North American fish fauna, *Ann. Miss. Bot. Gard.* **72**:596–635.

Wiley, E. O., Siegel-Causey, D., Brooks, D. R., and Funk, V. A., 1991, *The Compleat Cladist, Special Publication No. 19, The University of Kansas Museum of Natural History*, University of Kansas Press, Lawrence.

Willdenow, K., 1805, *Principles of Botany*, Blackwood, Edinburgh.

Williams, A., 1973, Distribution of brachiopod assemblages in relation to Ordovician palaeogeography, *Spec. Paps. Palaeontol.* **12**:241–269.

Williams, G. C., 1966, *Adaptation and Natural Selection*, Princeton University Press, Princeton.

Williams, G. C., 1992, *Natural Selection*, Oxford University Press, New York.

Williamson, P. G., 1981, Paleontological documentation of speciation in Cenozoic molluscs from Turkana Basin, *Nature* **293**:437–443.

Willis, J. C., and Yule, G. U., 1922, Some statistics of evolution and geographical distribution in plants and animals, and their significance, *Nature* **109**:177–179.

Wilson, E. O., 1988, *Biodiversity*, National Academy Press, Washington, D. C.

Wilson, E. O., 1993, *The Diversity of Life*, Harvard University Press, Cambridge.

Wilson, E. O., 1994, *Naturalist*, Island Press, Washington, D. C.

Wortman, J. L., 1903, Studies of Eocene Mammalia in the Marsh Collection, Peabody Museum: Primates, *Amer. J. Sci.* **15**:163–176, 399–414, 419–436.

Wright, S., 1931, Evolution in mendelian populations, *Genetics* **16**:97–159.

Zandee, M., and Roos, M. C., 1987, Component-compatibility in historical biogeography, *Cladistics* **3**:305–332.

Zink, R. M., 1991, The geography of mitochondrial DNA variation in two sympatric sparrows, *Evolution* **45**:329–339.

Index